Student Note-Taking Guide to accompany

SEVENTH EDITION

ENVIRONMENTAL SCIENCE

DANIEL D. CHIRAS

JONES AND BARTLETT PUBLISHERS

Sudbury, Massachusetts

BOSTON TORONTO LONDON SINGAPORE

World Headquarters

Jones and Bartlett Publishers
40 Tall Pine Drive
Sudbury, MA 01776
978-443-5000
info@jbpub.com
www.jbpub.com

Jones and Bartlett Publishers Canada
6339 Ormindale Way
Mississauga, ON L5V 1J2
CANADA

Jones and Bartlett Publishers International
Barb House, Barb Mews
London W6 7PA
UK

ISBN: 0-7637-4391-7

Printed in the United States of America
10 09 08 07 06 10 9 8 7 6 5 4 3 2 1

Contents

How This Book Can Help You Learn iv
Note-Taking Tips v
Chapter 1: Environmental Science and Critical Thinking 1
Chapter 2: Environmental Protection and Sustainable Development 7
Chapter 3: Understanding the Root Causes of the Environmental Crisis 13
Chapter 4: Principles of Ecology: How Ecosystems Work 19
Chapter 5: Principles of Ecology: Biomes and Aquatic Life Zones 27
Chapter 6: Principles of Ecology: Self-Sustaining Mechanisms in Ecosystems . . . 35
Chapter 7: Human Ecology: Our Changing Relationship with the Environment 45
Chapter 8: Population: Measuring Growth and Its Impact 49
Chapter 9: Stabilizing the Human Population: Strategies for Sustainability 57
Chapter 10: Creating a Sustainable System of Agriculture to Feed the World's People 63
Chapter 11: Preserving Biological Diversity 77
Chapter 12: Grasslands, Forests, and Wilderness: Sustainable Management Strategies 87
Chapter 13: Water Resources: Preserving Our Liquid Assets and Protecting Aquatic Ecosystems 97
Chapter 14: Nonrenewable Energy Sources 109
Chapter 15: Foundations of a Sustainable Energy System: Conservation and Renewable Energy 123
Chapter 16: The Earth and Its Mineral Resources 133
Chapter 17: Creating Sustainable Cities, Suburbs, and Towns: Principles and Practices of Sustainable Community Development 141
Chapter 18: Principles of Toxicology and Risk Assessment 149
Chapter 19: Air Pollution and Noise: Living and Working in a Healthy Environment 161
Chapter 20: Global Air Pollution: Ozone Depletion, Acid Deposition, and Global Warming 175
Chapter 21: Water Pollution: Sustainably Managing a Renewable Resource . . . 191
Chapter 22: Pest and Pesticides: Growing Crops Sustainably 203
Chapter 23: Hazardous and Solid Wastes: Sustainable Solutions 213
Chapter 24: Environmental Ethics: The Foundation of a Sustainable Society . . 227
Chapter 25: Sustainable Economics: Understanding the Economy and Challenges Facing the Industrial Nations 235
Chapter 26: Sustainable Economic Development: Challenges Facing the Developing Nations 249
Chapter 27: Law, Government, and Society 255
Photo Credits 267

How This Book Can Help You Learn

All of us have different learning styles. Some of us are visual learners, some more auditory, some learn better by doing an activity. Some students prefer to learn new material using visual aids. Some learn material better when they hear it in a lecture; others learn it better by reading it. Cognitive research has shown that no matter what your learning style, you will learn more if you are actively engaged in the learning process.

This Student Note-Taking Guide will help you learn by providing a structure to your notes and letting you utilize all of the learning styles mentioned above. Students don't need to copy down every word their professor says or recopy their entire textbook. Do the assigned reading, listen in lecture, follow the key points your instructor is making, and write down meaningful notes. After reading and lectures, review your notes and pull out the most important points.

The Student Note-Taking Guide is your partner and guide in note-taking. Your Guide provides you with a visual guide that follows the chapter topics presented in your textbook. If your instructor is using the PowerPoint slides that accompany the text, this guide will save you from having to write down everything that is on the slides. There is space provided for you to jot down the terms and concepts that you feel are most important to each lecture. By working with your Guide, you are seeing, hearing, writing, and, later, reading and reviewing. The more often you are exposed to the material, the better you will learn and understand it. Using different methods of exposure significantly increases your comprehension.

Your Guide is the perfect place to write down questions that you want to ask your professor later, interesting ideas that you want to discuss with your study group, or reminders to yourself to go back and study a certain concept again to make sure that you really got it.

Having organized notes is essential at exam time and when doing homework assignments. Your ability to easily locate the important concepts of a recent lecture will help you move along more rapidly, as you don't have to spend time rereading an entire chapter just to reinforce one point that you may not have quite understood.

Your Guide is a valuable resource. You've found a wonderful study partner!

Note-Taking Tips

1. It is easier to take notes if you are not hearing the information for the first time. Read the chapter or the material that is about to be discussed before class. This will help you to anticipate what will be said in class, and have an idea of what to write down. It will also help to read over your notes from the last class. This way you can avoid having to spend the first few minutes of class trying to remember where you left off last time.
2. Don't waste your time trying to write down everything that your professor says. Instead, listen closely and only write down the important points. Review these important points after class to help remind you of related points that were made during the lecture.
3. If the class discussion takes a spontaneous turn, pay attention and participate in the discussion. Only take notes on the conclusions that are relevant to the lecture.
4. Emphasize main points in your notes. You may want to use a highlighter, special notation (asterisks, exclamation points), format (circle, underline), or placement on the page (indented, bulleted). You will find that when you try to recall these points, you will be able to actually picture them on the page.
5. Be sure to copy down word-for-word specific formulas, laws, and theories.
6. Hearing something repeated, stressed, or summed up can be a signal that it is an important concept to understand.
7. Organize handouts, study guides, and exams in your notebook along with your lecture notes. It may be helpful to use a three-ring binder, so that you can insert pages wherever you need to.
8. When taking notes, you might find it helpful to leave a wide margin on all four sides of the page. Doing this allows you to note names, dates, definitions, etc., for easy access and studying later. It may also be helpful to make notes of questions you want to ask your professor about or research later, ideas or relationships that you want to explore more on your own, or concepts that you don't fully understand.
9. It is best to maintain a separate notebook for each class. Labeling and dating your notes can be helpful when you need to look up information from previous lectures.
10. Make your notes legible, and take notes directly in your notebook. Chances are you won't recopy them no matter how noble your intentions. Spend the time you would have spent recopying the notes studying them instead, drawing conclusions and making connections that you didn't have time for in class.
11. Look over your notes after class while the lecture is still fresh in your mind. Fix illegible items and clarify anything you don't understand. Do this again right before the next class.

Notes

Environmental Science SEVENTH EDITION — Daniel D. Chiras

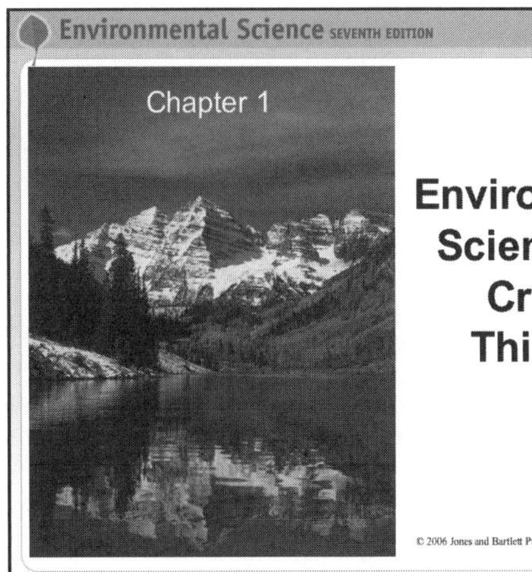

Chapter 1

Environmental Science and Critical Thinking

CH01-1

1.1 Encouraging Signs and Continuing Challenges

- Although there are some signs of improvement, the vast majority of environmental trends show signs of movement away from sustainability.
- Positive changes in perception and policy have led to real progress in solving local, regional, and global environmental problems. Unfortunately, the problems still outweigh the solutions.

CH01-2

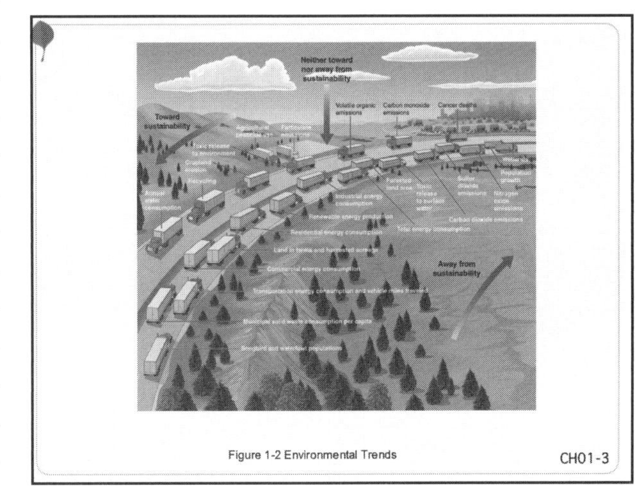

Figure 1-2 Environmental Trends

CH01-3

Notes

1.2 What is Environmental Science?

- It is a branch of science that seeks to understand how we affect our environment, and how these issues can be addressed.
- It is multidisciplinary, drawing from sciences and humanities.

CH01-4

1.3 Science and Scientific Method

- Science is a body of knowledge and a method of acquiring further knowledge about the world around us.

CH01-5

❖ Scientific Method: The Basis of Good Science

- Scientific knowledge is often gained through a deliberate process of observation and measurement.
- It leads scientists to formulate hypotheses that can then be tested through experiments.
- This is a form of **inductive reasoning.**

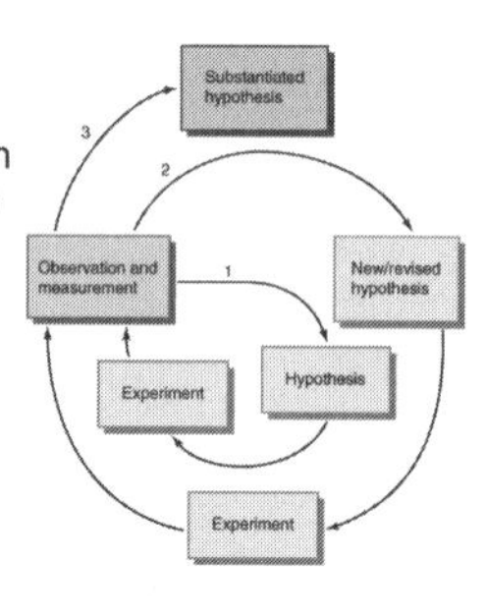

Figure 1-4 The scientific method

CH01-6

Notes

❖ Experimentation: Testing Hypotheses

- Experiments enable scientists to test hypotheses and gain new knowledge.
- Experiments must be carried out very carefully.
- Good experiments often require control and experimental groups.

Testing wheat response to elevated levels of atmospheric CO_2 (Courtesy of the USDA)

CH01-7

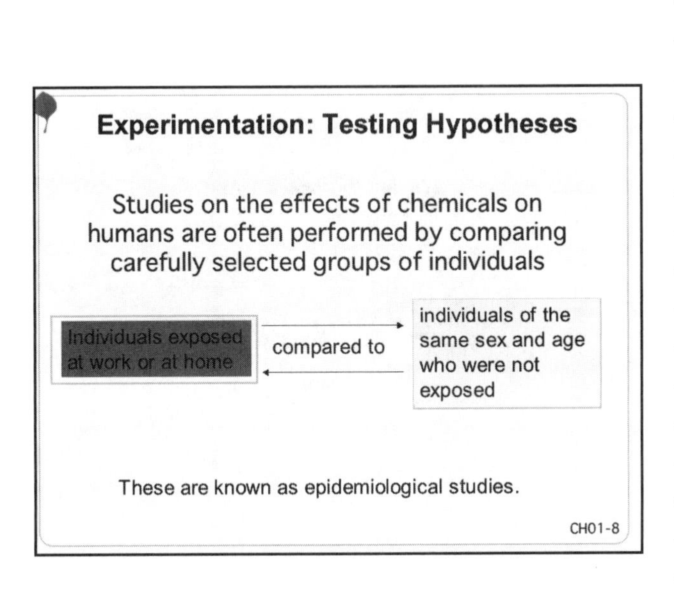

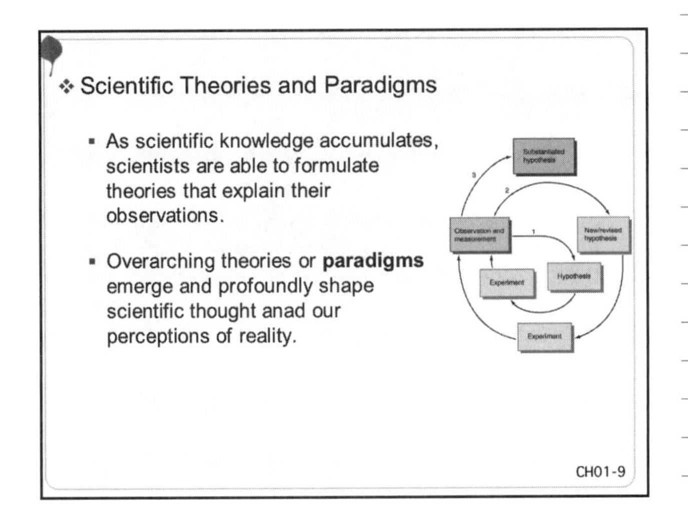

Notes

❖ Science and Values

- Scientists seek to be objective in their work so that science is fact-based and value-free.
- Biases do cloud objectivity at times. Nonetheless, scientific knowledge can also affect human values.

CH01-10

1.4 Critical Thinking Skills

- Critical thinking is an acquired skill that helps us:
  - analyze issues
  - discern the validity of experimental results and assertions

CH01-11

❖ **Gather All Information**

- Critical thinking requires one to know as much information about an issue as possible before rendering an opinion or making a decision.

CH01-12

❖ **Understand All Terms**

- To think critically about an issue, one must understand the terms and concepts related to it.

CH01-13

❖ **Question the Methods**

- Critical thinking requires that we know how information has been acquired and that we question the methods by which it was derived.

CH01-14

❖ **Question the Source**

- Critical thinking requires one to search for hidden biases and assumptions that may influence one's understanding of an issue or interpretation of data.

CH01-15

Notes

Notes

❖ Question the Conclusions
▪ Critical thinking requires us to question the conclusions drawn from facts to see if other interpretations might be possible.
CH01-16

❖ Tolerate Uncertainty
▪ Our knowledge of the world around us is evolving, so we must accept uncertainty and make decisions with the best information possible.
CH01-17

❖ Examine the Big Picture
▪ To become a critical thinker it is necessary to examine the big picture relationships and entire systems.
CH01-18

Chapter 2: Environmental Protection and Sustainable Development

Notes

Environmental Science SEVENTH EDITION

Daniel D. Chiras

Chapter 2

Environmental Protection and Sustainable Development

CH02-1

2.1 Environmental Protection and Sustainable Development

- Increasingly, environmental protection is being incorporated into all human actions and into the process of development.
- Meeting our needs while protecting the environment is called **sustainable development.**

CH02-2

❖ The Evolution of Sustainable Development

- Environmental protection has evolved from piecemeal local efforts to a much more comprehensive global strategy covering a wide assortment of environmental problems.
- This involves high levels of cooperation among states and nations

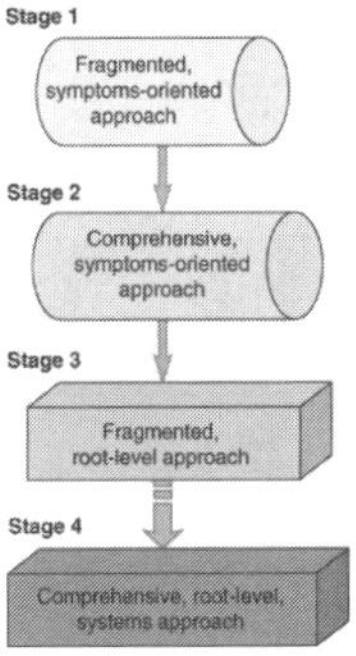

Fig. 2-1 The evolution of environmental protection

CH02-3

Notes

❖ The Next Generation of Environmental Protection Efforts

- Although environmental policy and protection efforts have evolved dramatically in the past three decades, most solutions dealt with symptoms.
- Efforts are now under way to address the root causes of the problems.

CH02-4

❖ Creating Sustainable Solutions

- All environmental problems result from the fact that human systems such as energy production and agriculture are unsustainable.
- They are inefficient in their use of resources, and most of them rely heavily on finite supplies of fossil fuels whose combustion creates many problems.

CH02-5

2.2 Meeting Human Needs while Protecting the Environment

❖ What Is Sustainable Development?

- Sustainable development is a means of meeting present needs without preventing future generations and other species from meeting their needs.
- Because the environment is essential to satisfying the needs of present and future generations, environmental protection is a key to its success.

CH02-6

❖ Satisfying the Triple Bottom Line

Sustainable development requires strategies that satisfy **social, economic, and environmental** goals simultaneously.

SUSTAINABLE DEVELOPMENT: CREATING A NEW RELATIONSHIP

Sustainable development calls for policies and actions that foster enduring relationships among people and between people and the planet

that result in the creation of

Sustainable communities, states, and nations,

which require

Social, economic, and environmental conditions conducive to harmony and long-term survival,

including

Social Conditions
- Adequate food, shelter, and other basic needs, creating a high quality of life
- Freedom from oppression
- Freedom from physical harm
- Democratically based decision making
- Participation
- Cooperation
- Intergenerational and intragenerational equity (environmental justice)

Environmental Conditions
- Waste output that does not exceed assimilative capacity of environment
- Use of renewable resources that does not exceed Earth's capacity to regenerate or renew them
- Use of nonrenewable resources that does not exceed the rate of their replacement by renewable substitutes
- Clean air and water
- Biodiversity

Economic Conditions
- Broad prosperity
- Greater local/regional self-reliance
- Ecologically sound economics
- Well-paying jobs for all who want them
- Healthy, stable economies

Fig. 2-3 Conditions created by sustainable development

CH02-7

❖ Principles of Sustainable Development

- Humans depend on the environment for countless goods and services that are essential to day-to-day living and the functioning of the economy.
- The renewable and nonrenewable resources that support our lives have very real limits, many of which we have exceeded.

Fig. 2-7

CH02-8

❖ Principles of Sustainable Development (cont.)

- Living sustainably means finding ways of prospering within limits.
- Intergenerational equity calls on us to live in ways that honor the needs of:
  - future generations
  - people alive today

CH02-9

Notes

Notes

❖ Principles of Sustainable Development (cont.)

- **Ecological justice** means that all species have a right to a clean environment and adequate resources.
- Building a sustainable society will require participation by and cooperation between governments, businesses, and individuals.
- To create a sustainable society, we must focus on strategies that address the root causes of environmental problems.

CH02-10

- Growth and Development: Understanding the Differences
  - Growth and development are fundamentally different goals
    - Growth results in an increase in material production and consumption that may be unsustainable in the long run.
    - Development is a strategy for improvements in culture that don't necessarily require further increases in resource consumption, pollution, and environmental destruction.

CH02-11

2.3 Human Settlements: Networks of (Unsustainable) Systems

- Human settlements are made up of many interdependent systems, such as:
  - Transportation
  - Energy
  - Waste management
- Growing evidence shows that these systems are not sustainable in the long run.

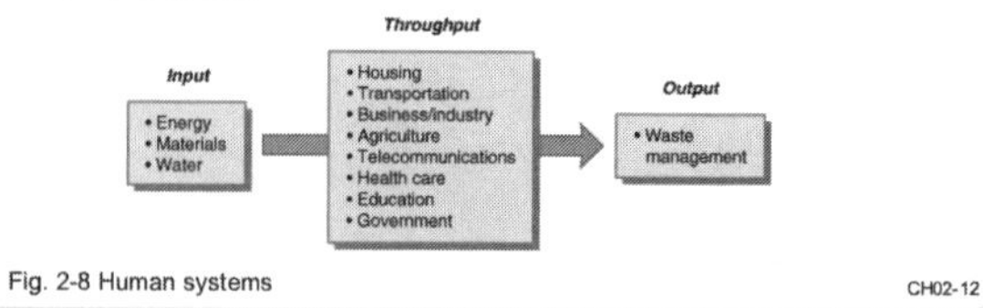

Fig. 2-8 Human systems

CH02-12

2.4 What's Wrong with Human Systems

- Although human systems may provide us with a steady stream of goods and services, they are systematically reducing the planet's carrying capacity.
- Human systems
  - (1) produce pollution in excess of the planet's ability to absorb and detoxify wastes,
  - (2) deplete nonrenewable resources faster than substitutes can be found, and
  - (3) use renewable resources faster than they can be regenerated.

CH02-13

❖ The Challenges Ahead

- To create a sustainable future, we must revamp the systems that support our lives.
- The challenge is twofold:
  - to retrofit existing systems to make them as sustainable as possible
  - to build new systems using the principles of sustainable design

CH02-14

2.5 Applying the Principles of Sustainable Development

❖ Operating Principles

- Creating a society that is sustainable requires a number of steps:
  - (1) population stabilization
  - (2) growth management
  - (3) efficiency
  - (4) renewable energy use
  - (5) recycling
  - (6) restoration
  - (7) sustainable resource management

CH02-15

Notes

Chapter 3: Understanding the Root Causes of the Environmental Crisis

Notes

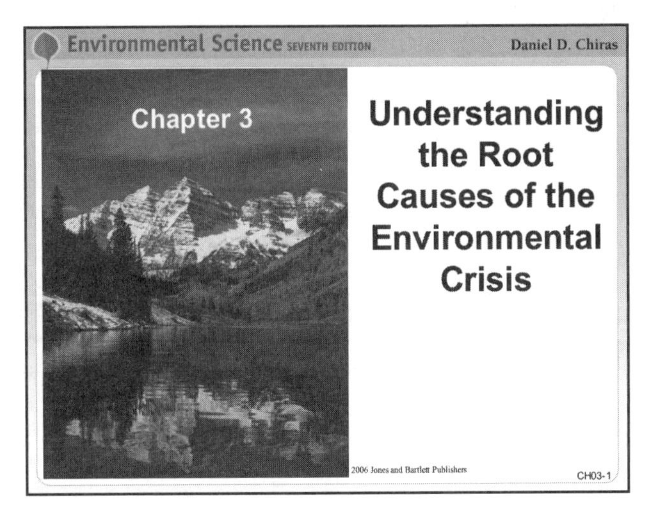

3.1 Roots of the Environmental Crisis

- Many different hypotheses have been presented to explain the root causes of the environmental crisis.
- When combined, these hypotheses do a good job of explaining environmental troubles.

CH03-2

❖ Religious Roots

- Some scholars think that early Christian teaching shaped many people's attitudes toward nature.
  - This fostered the creation of exploitive systems of science and technology that are largely responsible for the destruction of the environment.
- Others argue that although Christian teachings may have influenced thought, humans have a long history of environmental destruction going back long before the advent of Christianity.

CH03-3

Notes

❖ Some Cultural Roots: Democracy, Industrialization, and Frontierism

- Some scholars believe that the spread of democracy and the Industrial Revolution are at the root of the environmental crisis.
  - Democracy put land ownership and wealth in the hands of many
  - The Industrial Revolution made mass production of goods possible and spread wealth throughout society
- Frontierism, a belief in the inexhaustibility of resources, may have also been a root cause of the environmental crisis.

CH03-4

- Biological and Evolutionary Roots

- Human populations, like those of other organisms, expand if there are adequate supplies of resources and no other controls.
- The expansion of some organism populations is facilitated by special characteristics.
- For humans, technology has greatly facilitated population growth and increased our environmental impact.

CH03-5

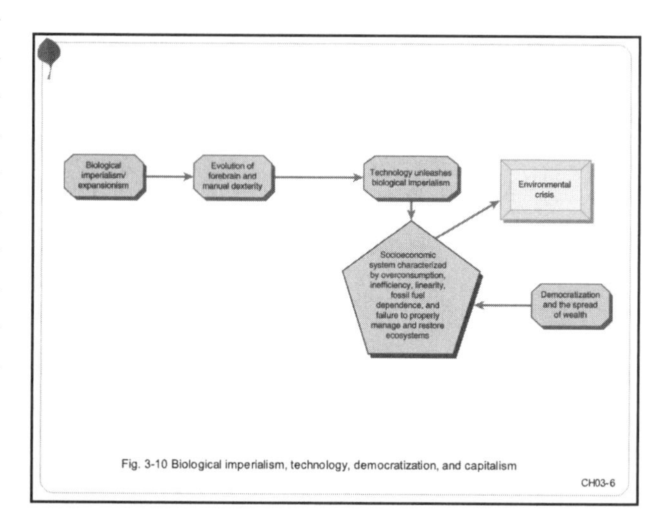

Fig. 3-10 Biological imperialism, technology, democratization, and capitalism

CH03-6

❖ Psychological and Economic Roots

- Human attitudes and beliefs are also responsible for many unsustainable practices.
- These factors, and others, influence our economic systems, laws, and way of life in profound ways.

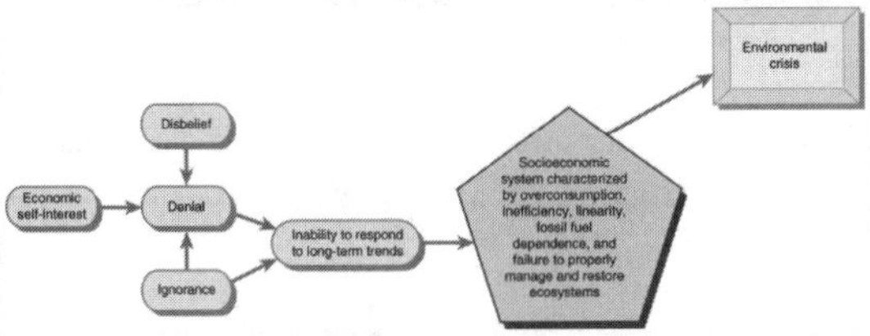

Environmental crisis

Disbelief

Economic self-interest

Denial

Inability to respond to long-term trends

Ignorance

Socioeconomic system characterized by overconsumption, inefficiency, linearity, fossil fuel dependence, and failure to properly manage and restore ecosystems

Fig. 3-11 Psychological factors

CH03-7

❖ Putting It All Together

- People, the systems that support them, and environmental issues are complex.
- So are the cause-and-effect relationships.

CH03-8

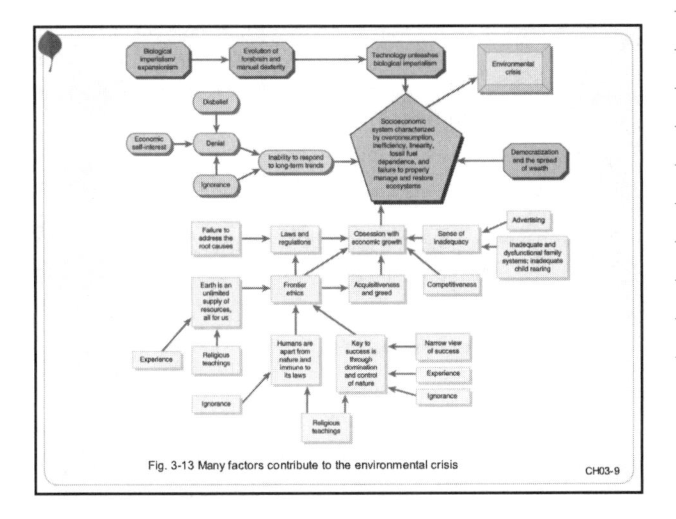

Fig. 3-13 Many factors contribute to the environmental crisis

CH03-9

Notes

Notes

3.2 Leverage Points

- Understanding the root causes of the environmental crisis helps us understand key **leverage points**.
- These points can be addressed to help put society back onto a sustainable course.

CH03-10

❖ Making Human Systems and Technologies Sustainable

- Unsustainable human systems are a result of many factors.
- Building a sustainable society will require a restructuring of basic human systems.
- Changes in technology will be essential to this transition.

CH03-11

❖ A New Worldview: Changing Our Perceptions, Values, and Beliefs

- Many observers believe that creating a sustainable society will require profound changes:
  - Changes in our understanding of issues, through education.
  - Changes in values.

CH03-12

❖ Unsustainable Ethics

- In most Western nations, human values express frontierism.
- This attitude causes us to pursue our own interests at the expense of the environment.

CH03-13

The Tenets of Frontierism

1. The Earth is an unlimited supply of resources for exclusive human use—there's always more, and it's all for us.
2. Humans are apart from nature, not a part of it. As a result, we often assume that we are immune to natural laws that govern other species.
3. Success comes through domination and control of nature.

CH03-14

- Frontier thinking influences how we design and operate all human systems, from the economy to government to education to waste management.
- Creating a sustainable society that protects the environment will require a new value system that:
  - respects limits
  - sees humans as a part of the natural world
  - recognizes the need to cooperate with nature, not dominate it

CH03-15

Notes

Notes

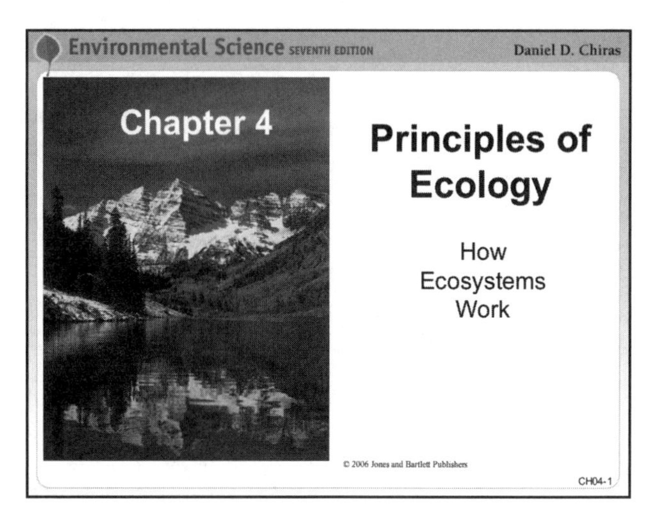

4.1 Humans and Nature: The Vital Connections

- Humans are a part of nature.
- We are dependent on natural systems for a variety of economically important resources and ecological services essential to our survival and long-term prosperity.

CH04-2

4.2 Ecology: The Study of Natural Systems

- Ecology is a field of science that seeks to describe relationships between organisms and their chemical and physical environment.

CH04-3

Notes

4.3 The Structure of Natural Systems

- The Biosphere
  - The biosphere is an enormous biological system, spanning the entire planet.
  - The materials within this closed system are recycled over and over in order for life to be sustained.
  - The only outside contribution to the biosphere is sunlight, which provides energy for all living things.

CH04-4

- Biomes and Aquatic Life Zones
  - The biosphere consists of distinct regions called biomes and aquatic life zones.
  - Each has its own chemical and physical conditions and unique assemblage of organisms.
  - Humans inhabit all biomes, but are most prevalent in those with the mildest climates.

CH04-5

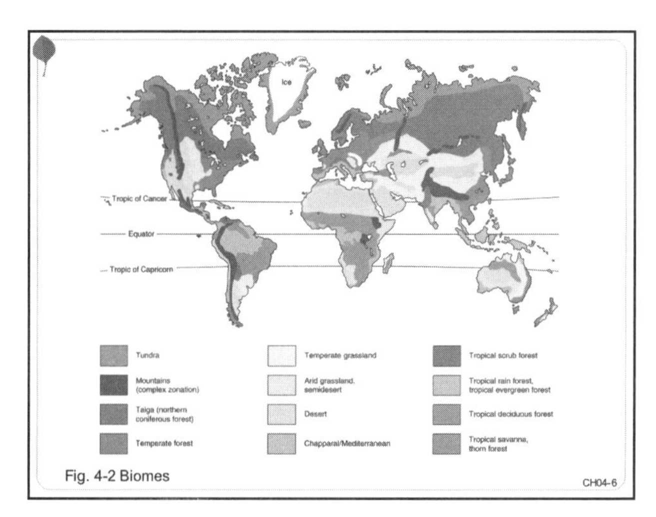

Fig. 4-2 Biomes

CH04-6

Notes

❖ What Is an Ecosystem?

- Ecosystems are biological systems consisting of organisms and their environment.
- Organisms thrive within a range of abiotic conditions.
  - Altering those conditions can have severe consequences and can even cause extinction.

CH04-8

❖ What Is an Ecosystem? (cont.)

- Organisms require many different abiotic factors to survive.
- One factor—the limiting factor—tends to be critical to survival and growth of a population.
- Altering concentrations of limiting factors can result in dramatic fluctuations in populations.

CH04-9

Notes

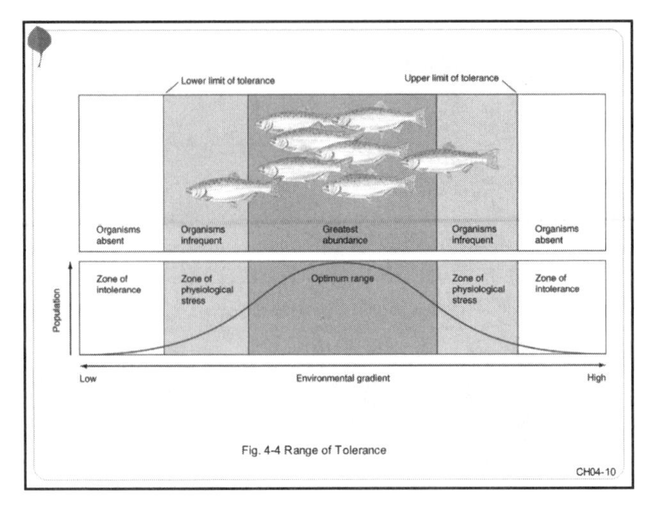

Fig. 4-4 Range of Tolerance

CH04-10

❖ What Is an Ecosystem? (cont.)

- Organisms are the biotic components of ecosystems; they form an interdependent community of life.
- Competition occurs between species occupying the same habitat if their niches overlap.
- Although competition is a naturally occurring process, natural systems have evolved to minimize it.

CH04-11

❖ What Is an Ecosystem? (cont.)

- Humans are a major competitive force in nature.
- Our advanced technologies and massive population size permit us to out-compete many species.
- Destroying other species through competition can be disadvantageous in the long run.

CH04-12

4.4 Ecosystem Function

- Photosynthetic organisms such as plants and algae produce food within ecosystems.
- Their well-being is essential to the survival and well-being of all other species.

CH04-13

❖ Food Chains and Food Webs

- Food and energy flow through food chains that are part of much larger food webs in ecosystems.

Fig. 4-7 Simplified grazer food chains

CH04-14

❖ The Flow of Energy and Nutrients through Food Webs

- Food chains are biological avenues for the flow of energy and the cycling of nutrients in the environment.
- Energy flows in one direction through food chains, but nutrients are recycled.

CH04-15

Notes

Notes

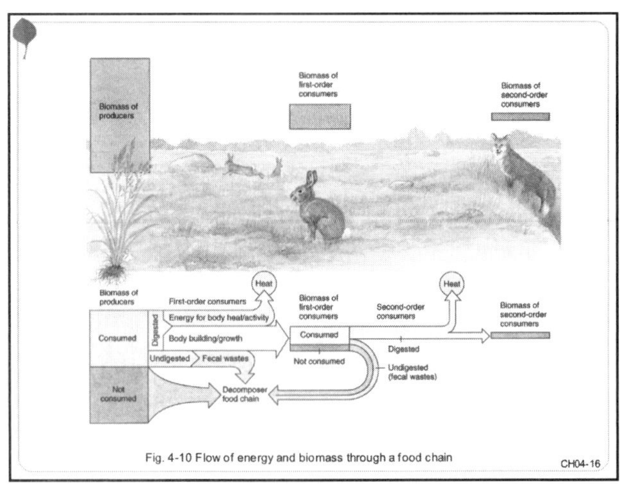

Fig. 4-10 Flow of energy and biomass through a food chain

CH04-16

❖ Trophic Levels

- The position of an organism in a food chain is called its trophic level.
    - Producers are on the first trophic level.
    - Herbivores are on the second level.
    - Carnivores are on the third level.
- The length of a food chain is limited by the loss of energy from one trophic level to another.
- The largest number of organisms is generally supported by the base of the food chain, the producers.

CH04-17

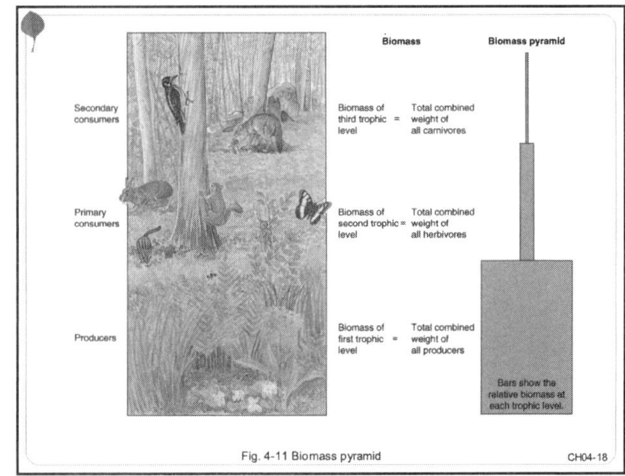

Fig. 4-11 Biomass pyramid

CH04-18

Notes

❖ Nutrient Cycles

- Nutrients are recycled in global nutrient cycles.
- In these cycles, nutrients alternate between organisms and the environment.
- Humans can disrupt nutrient cycles in many ways, with profound impacts on ecosystems.

Fig. 4-13 Nutrient cycle

CH04-19

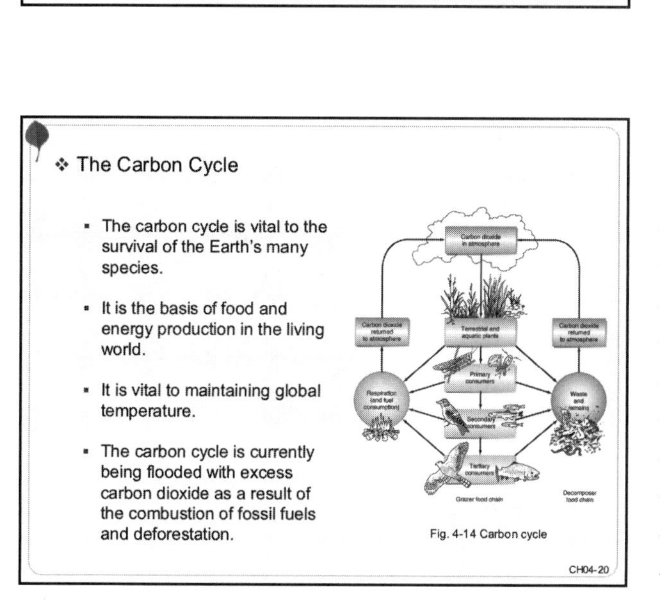

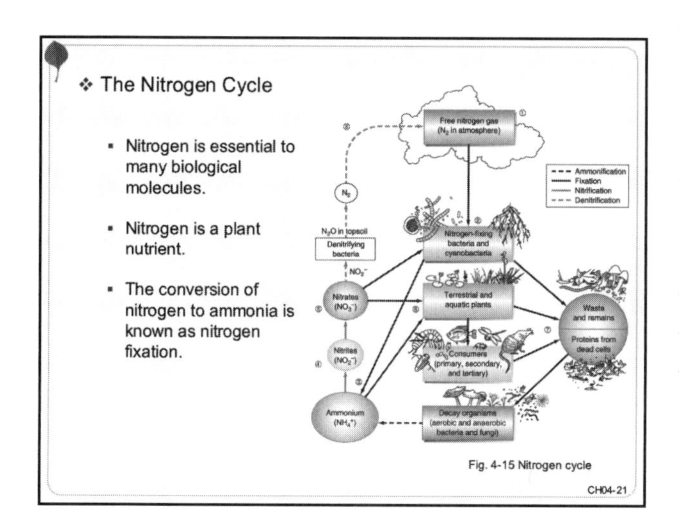

Notes

❖ The Nitrogen Cycle

- Humans alter the nitrogen cycle in at least four ways.

In the soil or water

- Applying excess nitrogen-containing fertilizer
- Disposing of nitrogen-rich municipal sewage
- Raising cattle in feedlots adjacent to waterways

In the atmosphere

- Burning fossil fuels, which releases nitrogen oxides

CH04-22

Chapter 5: Principles of Ecology: Biomes and Aquatic Life Zones

Notes

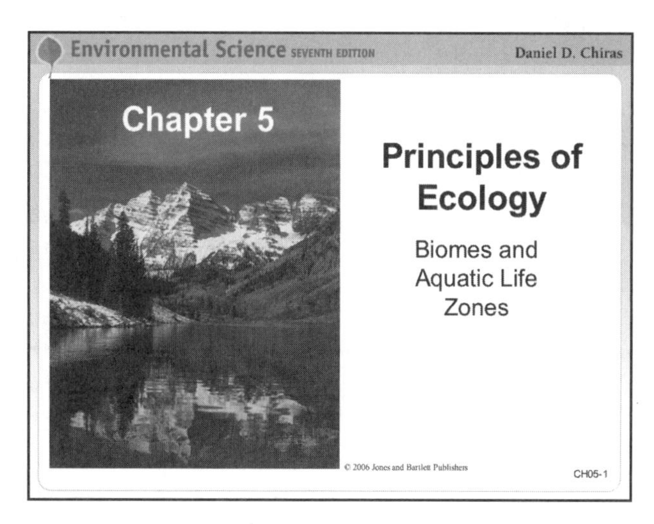

5.1 Weather and Climate: An Introduction

- Weather refers to daily conditions such as rainfall and temperature.
- Climate is the average weather over a long period.
- Climate determines the plant and animal life of a region.

CH05-2

❖ Major Factors That Determine Weather and Climate

- The Earth is unequally heated, which creates three major climatic zones:
  - Tropical
  - Temperate
  - Polar
- Air tends to flow from the equator to the poles.

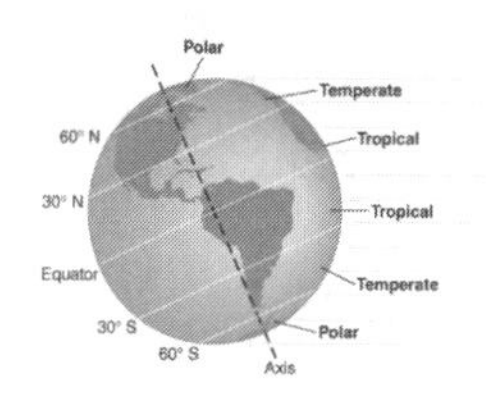

Fig. 5-1 Climate zones

CH05-3

Notes

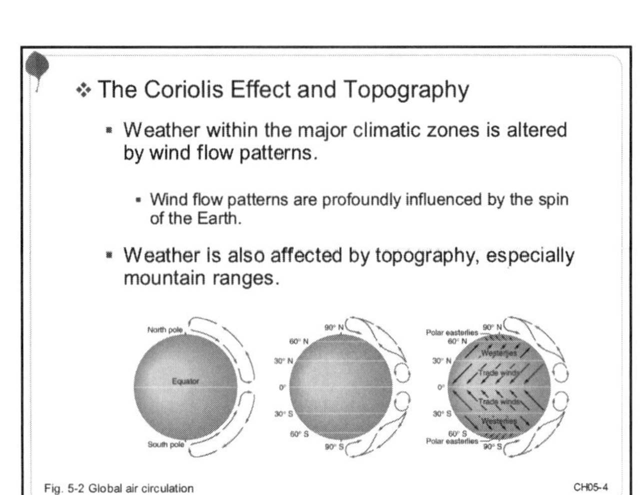

Fig. 5-2 Global air circulation

CH05-4

❖ Ocean Currents

- Warm water from the equator flows toward the poles.
- This warms landmasses near which it passes.

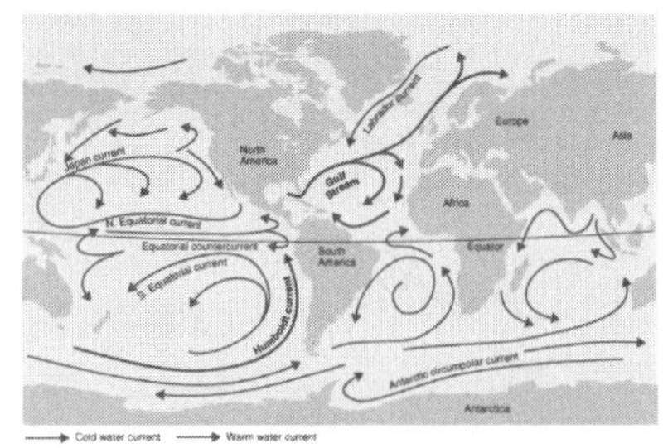

Fig. 5-4 Major ocean currents

CH05-5

5.2 The Biomes

- The Earth's surface can be divided into biologically distinct zones called biomes.
- Each has a distinct climate and unique assemblage of plants and animals.
- Regional variations occur within each biome.

CH05-6

Tundra
Mountains (complex zonation)
Taiga (northern coniferous forest)
Temperate forest
Temperate grassland
Arid grassland, semidesert
Desert
Chapparal/Mediterranean
Tropical scrub forest
Tropical rain forest, tropical evergreen forest
Tropical deciduous forest
Tropical savanna, thorn forest

Fig. 5-1 The Earth's biomes

CH05-7

Notes

- The Tundra
  - The Arctic tundra, the northernmost biome, is characterized by the harshest climate.
  - Because the growing season is so short, life on the Arctic tundra is extremely vulnerable to human actions.

CH05-8

- The Taiga
  - The taiga is a band of coniferous trees spreading across the northern continents south of the tundra.
  - Its climate is milder and its life-forms are more diverse than those of the tundra.
  - The taiga supports many large populations of wild animals.
  - The forests of this region are under heavy pressure to meet rising demands for wood and wood products.

Notes

The Temperate Deciduous Forest Biome

- The temperate deciduous forest biome occurs in regions with abundant rainfall and long growing seasons.
- This biome has been heavily settled by humans and dramatically altered.

CH05-10

The Grassland Biome

- The grassland biome occurs in regions of intermediate precipitation—enough to support grasses but not enough to support trees.
- On most continents, the rich soil of the biome has been heavily exploited by humans for agriculture.

CH05-11

The Desert Biome

- The desert biome is characterized by dry, hot conditions.

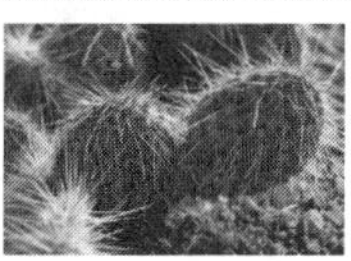

- Often, it abounds with plants and animals adapted to the heat and lack of moisture.
- The world's deserts are expanding because of human activities such as overgrazing livestock and the production of greenhouse gases.

CH05-12

Notes

The Tropical Rain Forest Biome

- The tropical rain forest is the richest and most diverse biome on Earth because of its abundant rainfall and warm climate.
- About half of the world's rain forest has been destroyed.
- Huge tracts could be eliminated in the near future, with devastating effects on climate, plants, and wildlife if current trends continue.

CH05-13

Altitudinal Biomes

- Because climate varies with altitude, the distribution and abundance of life also change.

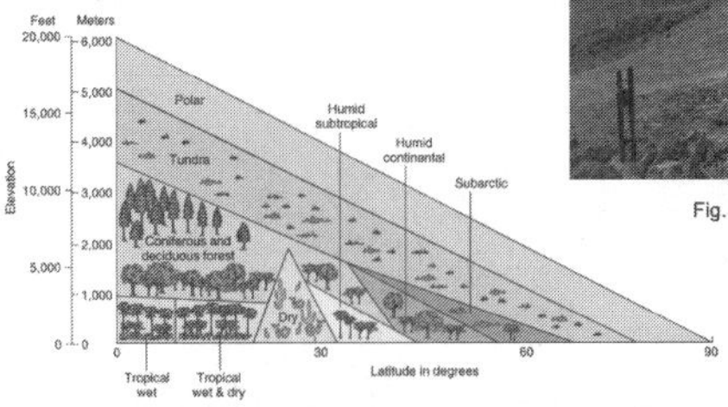

Fig. 5-20 Alpine tundra

Fig. 5-19 Altitudinal biomes

CH05-14

5.3 Aquatic Life Zones

- Aquatic systems are divided into distinct regions, known as aquatic life zones.
- They may be freshwater or saltwater.
- The abundance of life is determined by energy and nutrient levels.
- Phytoplankton form the base of aquatic food chains.

CH05-15

Notes

❖ Freshwater Lakes

- Lakes are divided into four regions:
  - the littoral zone
  - the limnetic zone
  - the profundal zone
  - the benthic zone
- Each has very different conditions and, consequently, very different life-forms.

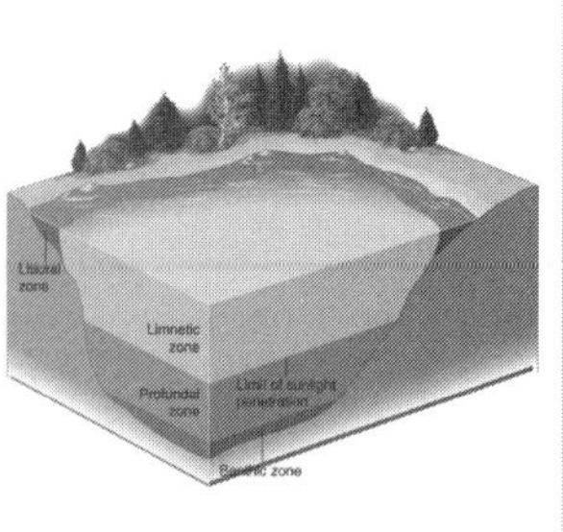

Fig. 5-21 Zones of a lake

CH05-16

❖ Rivers and Streams

- Rivers and streams are complex ecosystems that rely more on agitation for oxygenation of their waters than lakes do.
- Many nutrients in streams that support aquatic life come from neighboring terrestrial ecosystems.
- The quality of water in a stream is profoundly influenced by activities in the watershed.

CH05-17

❖ Protecting Freshwater Ecosystems

- Like lakes and ponds, streams are self-purging, but extremely vulnerable to pollution if sources exceed the capacity to self-cleanse.

Acid drainage precipitating iron hydroxide.

CH05-18

Notes

❖ Saltwater Life Zones

- The oceans cover over 70% of the Earth's surface
- The oceans can be divided into ecologically distinct life zones

CH05-19

❖ The Coastal Life Zones

- The coastlines are highly productive waters.
- They are characterized by abundant sunlight and a rich supply of nutrients, both of which contribute to an abundance of life-forms.

Fig. 5.25 Coastal wetland

CH05-20

❖ The Coastal Life Zones

- Estuaries are nutrient-rich zones at the mouths of rivers.
- They are often associated with coastal wetlands, together forming the estuarine zone.
- The estuarine zone is highly productive and of great value to humans and other species.
- Human activities severely threaten this important biological asset.

CH05-21

Notes

❖ The Coastal Life Zones

- The shorelines of the world are rocky or sandy regions.
- They are home to a surprising number of organisms adapted to the tides and the turbulence created by wave action.
- Coral reefs are the aquatic equivalent of the tropical rain forests and are being rapidly destroyed.

CH05-22

❖ The Marine Ecosystem

- The marine ecosystem consists of four ecologically distinct life zones, similar to those found in lakes.

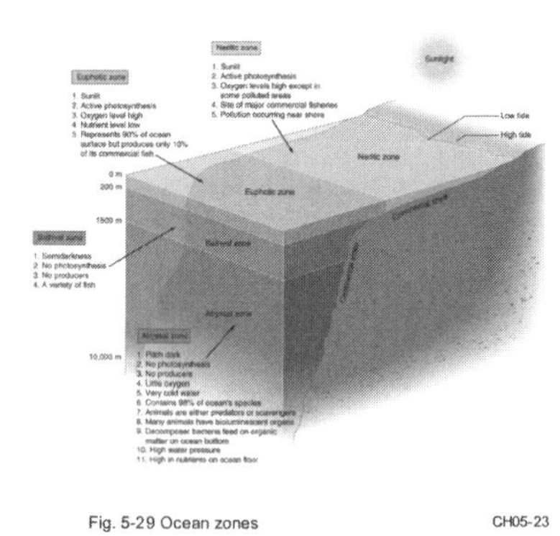

Fig. 5-29 Ocean zones

CH05-23

Chapter 6: Principles of Ecology: Self-Sustaining Mechanisms in Ecosystems

Notes

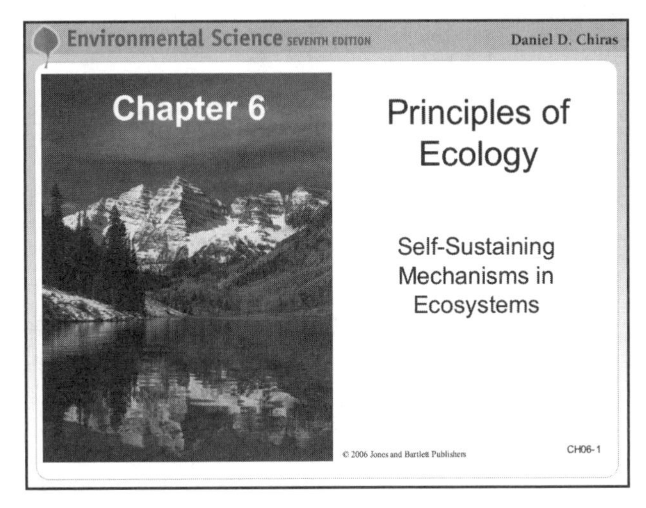

6.1 Homeostasis: Maintaining the Balance

- Homeostasis is a state of relative constancy vital to the survival of organisms.

CH06-2

❖ Homeostasis in Natural Systems

- Like organisms, ecosystems possess many mechanisms that either resist change or help them recover from change.
- These mechanisms help keep natural systems in a state of relative constancy.

CH06-3

Notes

❖ Factors That Contribute to Ecosystem Homeostasis

- Numerous biotic and abiotic factors influence the growth of populations.
- Some stimulate growth; others deter growth.
- Ecosystem homeostasis is the result of the interaction of these factors.

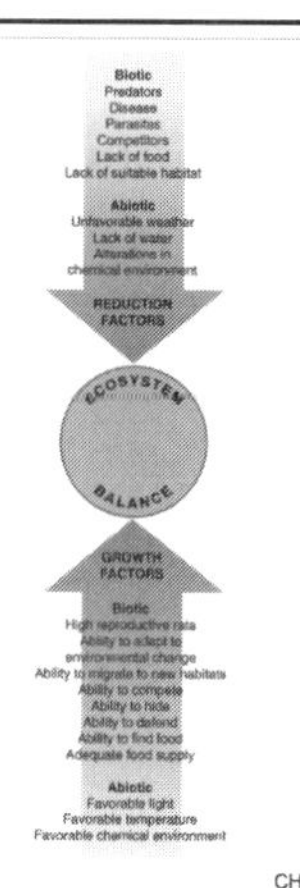

Fig. 6-2. Ecosystem balance

CH06-4

❖ The Resilience of Ecosystems

- In ecosystems, changes in biotic and abiotic conditions lead to a cascade of effects, but the systems tend to return to normal over time.
- The ability to resist change is called resilience.

CH06-5

❖ Resisting Changes from Human Activities

- The ability of ecosystems to recover from small changes minimizes and sometimes negates the impacts of human actions.
- In many instances human actions can overwhelm the recuperative capacity of natural systems.

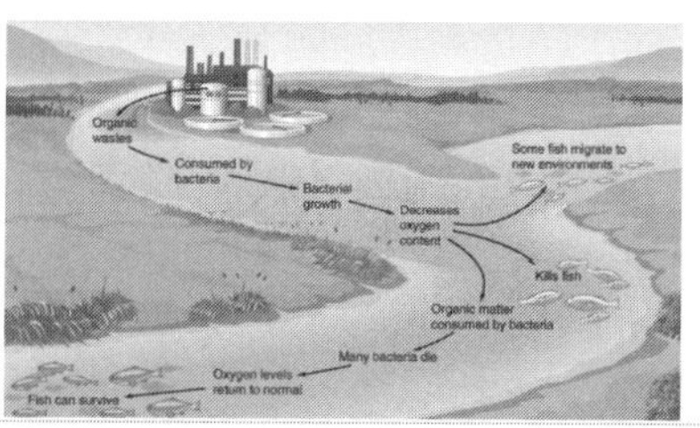

Fig. 6-4 Regaining balance

CH06-6

Notes

❖ Population Control and Sustainability

- Natural systems are sustained in large part by intrinsic and extrinsic mechanisms that help to maintain populations within the carrying capacity of the environment.
- To create a sustainable society, many experts believe that humans must also control population size.

CH06-7

❖ Species Diversity and Stability

- Ecologists still debate the reasons why some ecosystems are stable and some are not.
- We do know that reductions in species diversity can destabilize ecosystems.

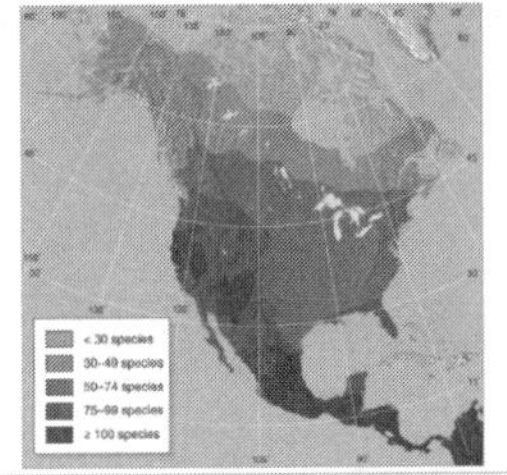

Fig. 6-6 Diversity vs. latitude

CH06-8

6.2 Natural Succession: Establishing Life on Lifeless Ground

- Ecosystems can form on barren or relatively lifeless ground by a process called natural succession.

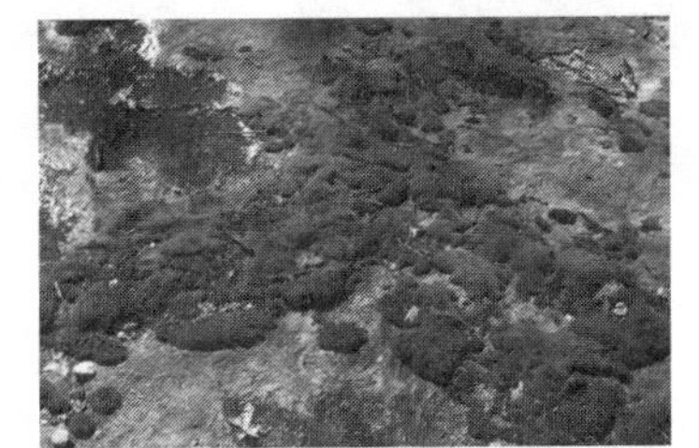

CH06-9

Notes

❖ Primary Succession: Starting from Scratch

- Primary succession is a process in which mature ecosystems form on barren ground where none previously existed.

Fig. 6-7b Primary succession

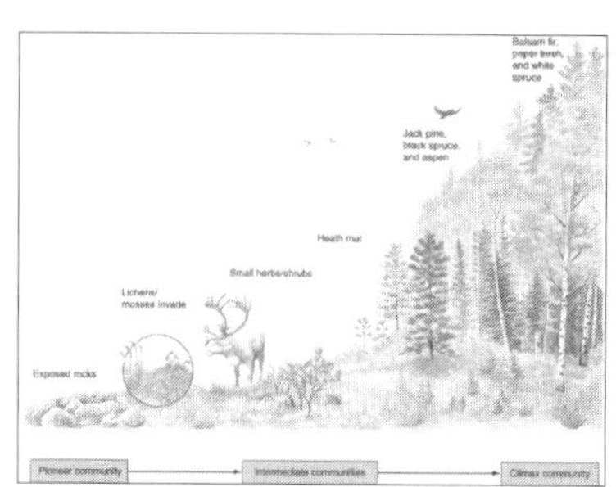

CH06-10

❖ Secondary Succession: Natural Ecosystem Restoration

- Secondary succession is a long-term repair process that takes place after an ecosystem is destroyed by natural or human causes.
- It occurs more rapidly than primary succession because there's generally no need to form soil.

Seedling in Mt. St. Helens ash. (U.S. Army Corps of Engineers)

Mt. St. Helens, one year after its eruption.

Fig. 6-8 Mt. St. Helens, twenty-five years after its eruption.

❖ Changes During Succession: An Overview

- During succession, biotic and abiotic conditions change.
- Pioneer and intermediate communities alter conditions so much that they promote the growth of new communities that eventually replace them.
- During succession, two of the most notable changes are:
  - an increase in species diversity
  - ecosystem stability

CH06-13

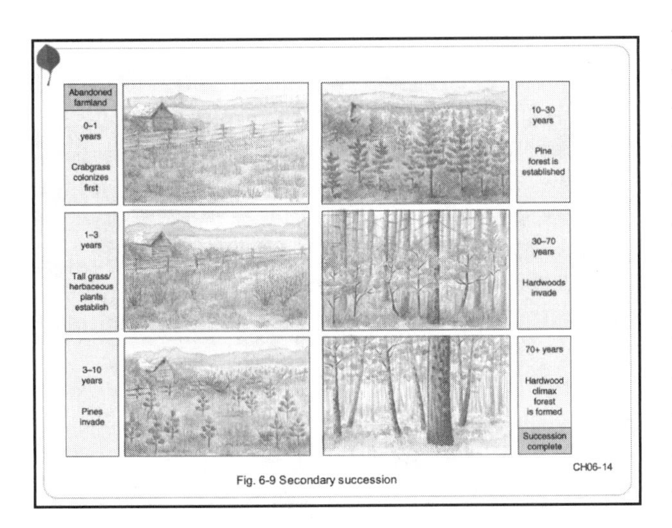

Fig. 6-9 Secondary succession

CH06-14

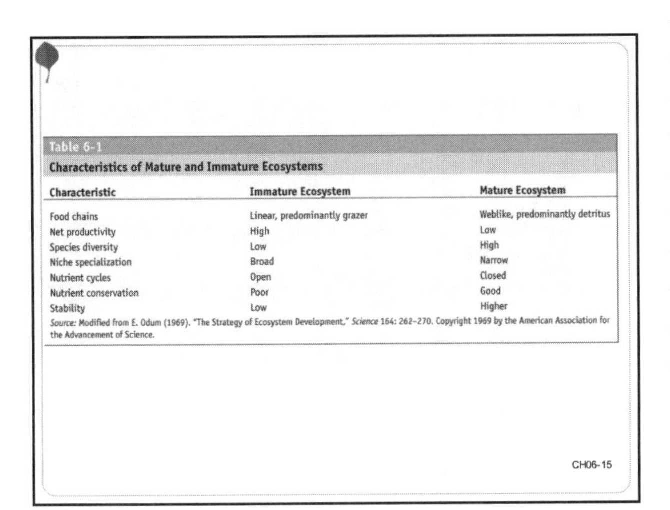

Table 6-1

Characteristics of Mature and Immature Ecosystems

Characteristic	Immature Ecosystem	Mature Ecosystem
Food chains	Linear, predominantly grazer	Weblike, predominantly detritus
Net productivity	High	Low
Species diversity	Low	High
Niche specialization	Broad	Narrow
Nutrient cycles	Open	Closed
Nutrient conservation	Poor	Good
Stability	Low	Higher

Source: Modified from E. Odum (1969). "The Strategy of Ecosystem Development," *Science* 164: 262–270. Copyright 1969 by the American Association for the Advancement of Science.

CH06-15

Notes

Notes

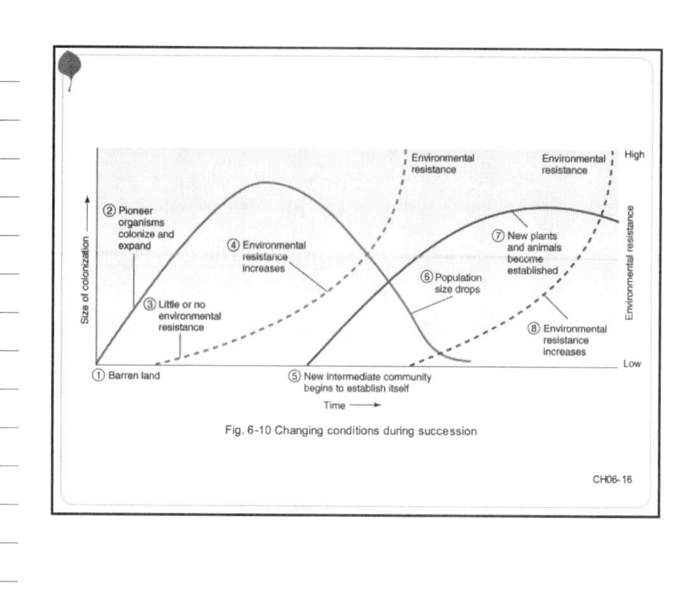

Fig. 6-10 Changing conditions during succession

CH06-16

6.3 Evolution: Responding to Change

- The ability of species to evolve in response to changes in biotic and abiotic conditions has contributed to the sustainability of life on Earth.

CH06-17

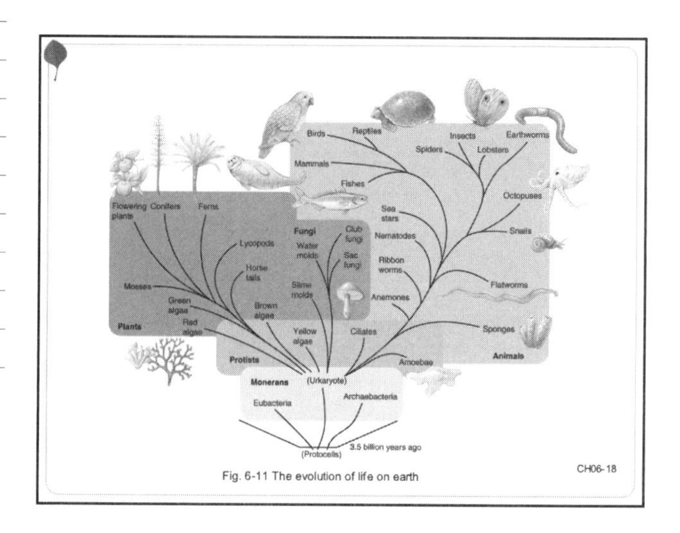

Fig. 6-11 The evolution of life on earth

CH06-18

Notes

❖ Evolution by Natural Selection: An Overview

- Natural selection is the driving force behind evolution.
- It consists of natural forces that select for those members of a population that are superior in one or more features.
- These advantages increase chances of surviving and reproducing.

CH06-19

❖ Genetic Variation: The Raw Material of Evolution

- Genetic differences in organisms of a population are called genetic variation.
- Genetic variation of organisms in a population may give some members of a population an advantage over others.

CH06-20

❖ Genetic Variation: The Raw Material of Evolution

- Genetic variation comes from:
  - mutations
  - sexual reproduction
  - "crossing over," a process which occurs during the formation of sperm and ova
- Genetic variation produces variation in populations in structure, function, and behavior.

CH06-21

Notes

Natural Selection: Nature's Editor

- Natural selection weeds out the less fit organisms of a population, leaving behind the fittest.

CH06-22

❖ Speciation: How New Species Form

- Evolution may result in the formation of new species.
- When members of a population are separated by a physical barrier, they may evolve along different lines, forming separate species.

CH06-23

❖ Coevolution and Ecosystem Balance

- Organisms can evolve in concert with one another.
- Changes in one organism result in changes in the other.
- This process is called **coevolution**.

CH06-24

6.4 Human Impact on Ecosystems

- Human activities alter the environment by changing its biotic and abiotic components—direct ly or indirectly.

CH06-25

Notes

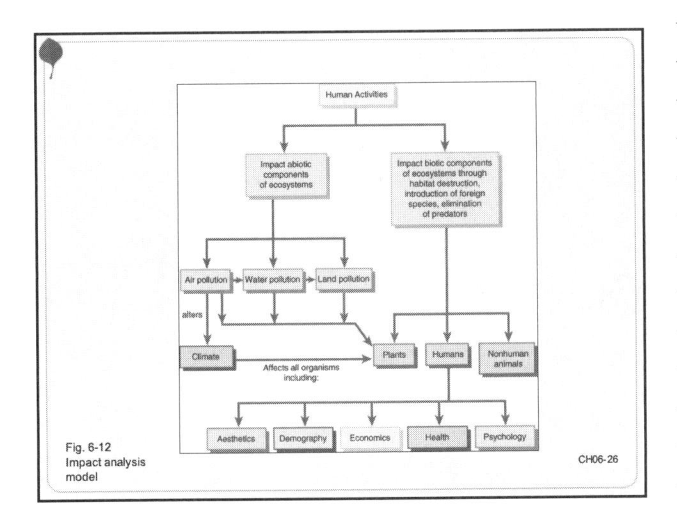

Fig. 6-12
Impact analysis model

CH06-26

- Altering Abiotic Factors
  - Human activities often alter the chemical and physical nature of the environment (the abiotic conditions).
  - This can affect us and the species that share this planet with us.

CH06-27

Notes

❖ Altering Biotic Factors

- Many human activities have a direct effect on the biotic components of ecosystems.
- Introduction of foreign species is particularly troublesome.

Fig. 6-13 Zebra mussels. (Courtesy of U.S. Fish & Wildlife Services)

- These species may proliferate without control, causing major economic and environmental damage.

CH06-28

❖ Simplifying Ecosystems

- Tampering with abiotic and biotic factors tends to reduce species diversity and thus simplify ecosystems.
- This makes them considerably more vulnerable to natural forces.

CH06-29

❖ Why Study Impacts?

- Being able to predict impacts permits us to select the least harmful and most sustainable development options.

Fig. SOSD 6-1
Restored wetland

CH06-30

Chapter 7: Human Ecology: Our Changing Relationship with the Environment

Notes

Environmental Science SEVENTH EDITION

Daniel D. Chiras

Chapter 7

Human Ecology

Our Changing Relationship with the Environment

CH07-1

7.1 Human Biological Evolution: Some Highlights

- Humans evolved from primates.
- We have evolved remarkable abilities that give us a distinct advantage over many other species.
- These advantages have permitted us to colonize much of the world and radically reshape the environment, sometimes to our own detriment.

Fig. 7-1 Evolution of human beings

CH07-2

7.2 Human Cultural Evolution: Our Changing Relationship with Nature

- Humans have evolved culturally through three distinct phases:
  1. hunting and gathering
  2. agricultural
  3. industrial
- During this time, our interaction with the environment and our impact on it have shifted dramatically away from sustainability.

CH07-3

Notes

❖ Hunting and Gathering Societies

- Throughout most of human evolutionary history, our ancestors were hunter-gatherers who lived in relative harmony with the Earth.
- Three main features of these societies held human impact to sustainable levels:
  - primitive technology
  - small population size
  - nomadic lifestyle

CH07-4

❖ Agricultural Societies

- Agriculture started as a subsistence activity.
- The plow and other forms of technology gave our ancestors the ability to produce excess food.
- This displaced farm workers, who took up crafts and trades in cities and towns, and fostered an upsurge in human population.

CH07-5

❖ The Industrial Society

- The Industrial Revolution provided another boost to human civilization, increasing:
  - population size
  - resource demand
  - pollution
  - environmental destruction
- It dramatically altered the human–environment interaction.
  - Attitudes toward nature shifted even farther away from stewardship.

Fig. 7-5 Industrialization on the farm.

CH07-6

Notes

❖ The Advanced Industrial Age

- In recent times, industrialization has grown rapidly.
- Resource demand and environmental destruction have reached unsustainable levels.

CH07-7

7.3 The Population, Resources, and Pollution Model

❖ Human populations acquire and use resources from the environment to produce goods and services.

❖ These activities degrade the environment by altering abiotic and biotic conditions.

CH07-8

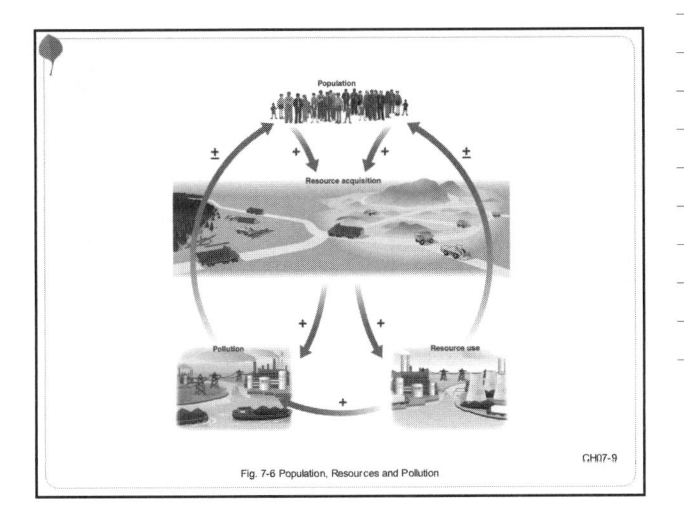

CH07-9

Fig. 7-6 Population, Resources and Pollution

Notes

7.4 The Sustainable Society: The Next Step

- Many steps are under way to create a more enduring way of life.
- These changes may be the early signs of a new cultural shift—a Sustainable Revolution.
- They are designed to restructure human systems to honor the limits of the natural systems on which all living things depend.

CH07-10

Notes

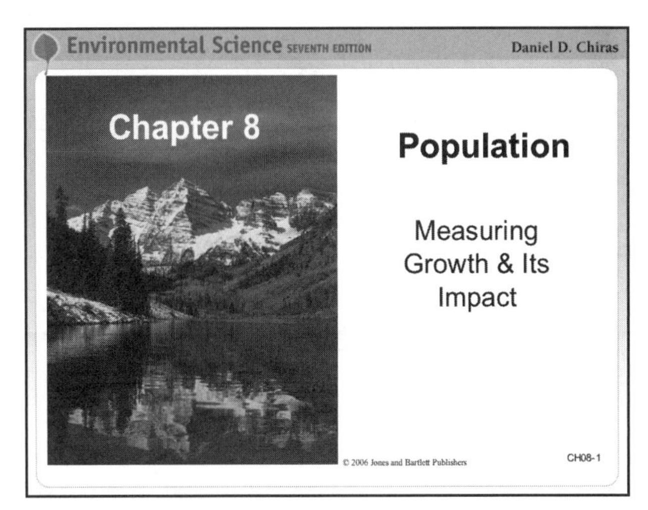

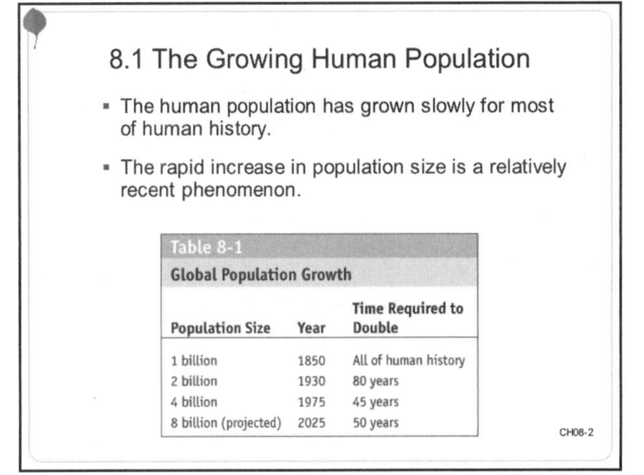

Table 8-1
Global Population Growth

Population Size	Year	Time Required to Double
1 billion	1850	All of human history
2 billion	1930	80 years
4 billion	1975	45 years
8 billion (projected)	2025	50 years

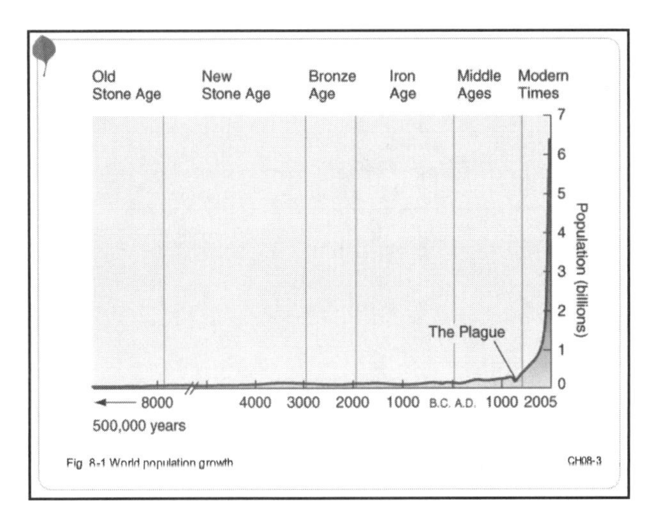

Fig 8-1 World population growth

Notes

❖ Why Has the Human Population Grown So Large?

- The human population has skyrocketed in the last 200 years.
- This is primarily because of a worldwide lowering of death rates without a corresponding decrease in birth rates.
- Death rates plummeted primarily as a result of increases in food supply and better medicine and sanitation.

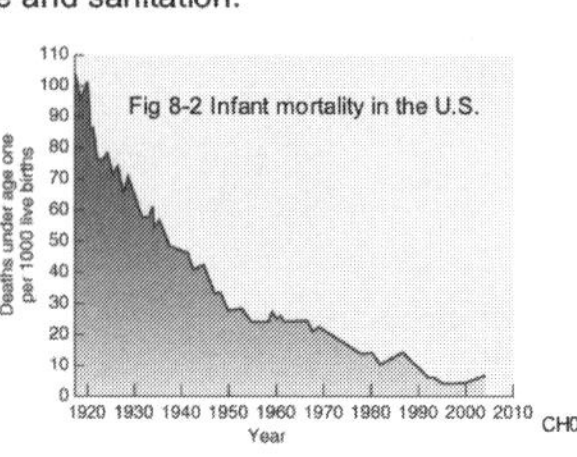

CH08-4

❖ Expanding the Earth's Carrying Capacity: An Ecological Perspective

- Technological advances lower environmental resistance and promote population growth.
- This increases the carrying capacity for humans.
- This, in turn, decreases the prospects of other species and may cause adverse effects in human populations as well.

CH08-5

❖ What Is the Earth's Carrying Capacity for Humans?

- Determining the Earth's carrying capacity for humans is a task fraught with difficulty.
- Some people think we have not reached the carrying capacity.
- Others believe that the human population already exceeds the Earth's long-term carrying capacity.

CH08-6

Notes

❖ Too Many People, Reproducing Too Quickly

- Population is at the root of virtually all environmental problems, including pollution and resource depletion, as well as many social and economic problems.
- The size of the population and the rate of growth both have significant impacts on environmental problems and solutions.

CH08-7

❖ Too Many People, Reproducing Too Quickly

- The massive size of the human population causes environmental problems evident in urban and rural areas.
- These include:
  - shortages of resources
  - environmental deterioration
  - a host of possible social problems

CH08-8

❖ Reproducing Too Quickly

- Social, economic, and environmental problems of cities and rural areas are aggravated by rapid population growth.
- The large size of many populations makes it difficult for governments to keep up with current demands.
- Continued rapid growth makes it nearly impossible to improve conditions and create a sustainable human presence.

CH08-9

Notes

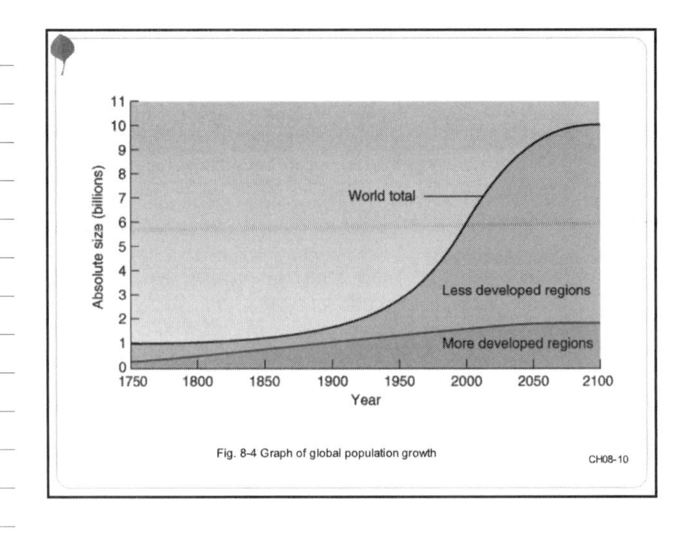

Fig. 8-4 Graph of global population growth

CH08-10

8.2 Understanding Populations and Population Growth

- Measuring Population Growth – Growth Rates and Death Rates
  - Global population growth is determined by subtracting the crude death rate from the crude birth rate.

CH08-11

- Doubling Time
  - Doubling time is the time it takes a population to double in size.
  - It is determined by dividing 70 by the growth rate.
  - Even relatively small growth rates result in rapid doubling.
  - Growth rates in the more developed countries are relatively low.
  - In the less developed nations, growth rates are generally much higher.

Table 8-2
Growth Rate and Doubling Time

Region	Growth Rate (%)	Doubling Time (years)*
World	1.2	58
More Developed Countries	0.1	700
Less Developed Countries	1.5	47
Africa	2.4	29
Asia	1.3	54
North America	0.5	140
Latin America	1.6	44
Europe	−0.2	—
Oceania	1.0	70

*Discrepancies in doubling time are the result of rounding off growth rates.

CH08-12

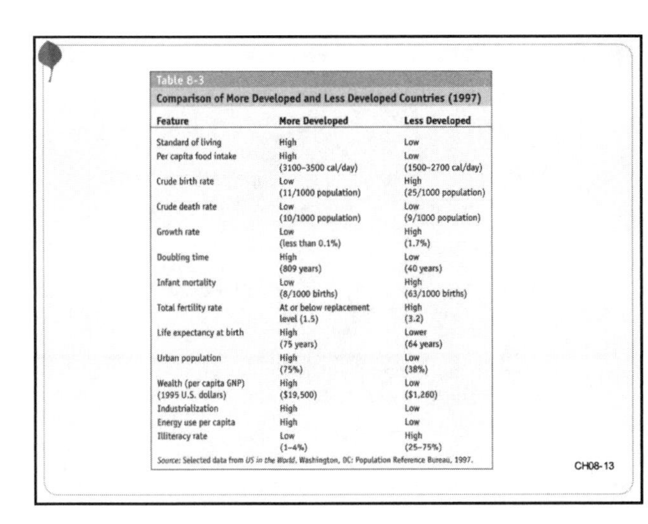

Table 8-3

Comparison of More Developed and Less Developed Countries (1997)

Feature	More Developed	Less Developed
Standard of living	High	Low
Per capita food intake	High (3100–3500 cal/day)	Low (1500–2700 cal/day)
Crude birth rate	Low (11/1000 population)	High (25/1000 population)
Crude death rate	Low (10/1000 population)	Low (9/1000 population)
Growth rate	Low (less than 0.1%)	High (1.7%)
Doubling time	High (809 years)	Low (40 years)
Infant mortality	Low (8/1000 births)	High (63/1000 births)
Total fertility rate	At or below replacement level (1.5)	High (3.2)
Life expectancy at birth	High (75 years)	Lower (64 years)
Urban population	High (75%)	Low (38%)
Wealth (per capita GNP) (1995 U.S. dollars)	High ($19,500)	Low ($1,260)
Industrialization	High	Low
Energy use per capita	High	Low
Illiteracy rate	Low (1–4%)	High (25–75%)

Source: Selected data from *US in the World*. Washington, DC: Population Reference Bureau, 1997.

CH08-13

❖ The Total Fertility Rate and Replacement-Level Fertility

- **Total fertility rate** is the average number of children women are expected to have during their reproductive age span.
- **Replacement-level fertility** occurs when couples produce exactly the number of children needed to replace themselves.
- **Zero population growth** occurs when the death rate equals the birth rate and when the net migration is zero.

CH08-14

❖ Migration

- The growth of a town, city, state, or region is determined by two factors:
  - Growth rate (natural increase)
  - Migration (the movement of people into and out of the population)
- Immigration, the movement of people from one region of a country to another, affects regional population growth.

CH08-15

Notes

Notes

❖ Population Histograms

- Population histograms are graphs of populations that depict the various age groups for males and females and provide useful information for planners.
- The histograms of countries may be expansive, constrictive, or stationary.
- Overall, the world population is expansive, in large part because of continued growth in the less developed nations.

CH08-16

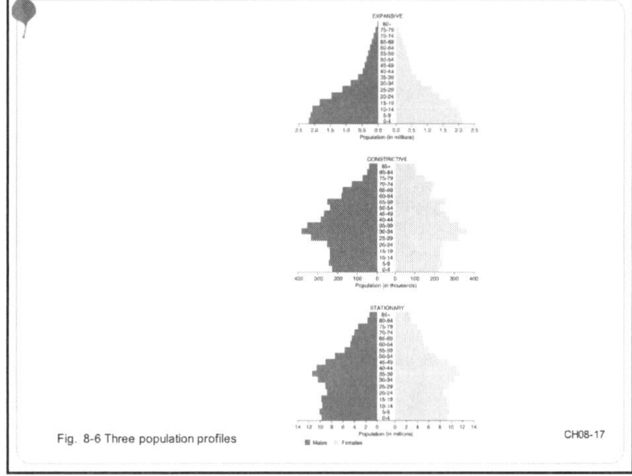

Fig. 8-6 Three population profiles

CH08-17

❖ Exponential Growth

- Human populations are growing exponentially.
- Globally, the human population has "rounded the bend" of the exponential growth curve.
- This means that even small percentage increases result in huge numbers of new world residents.

CH08-18

❖ Exponential Growth

- Exponential growth is cause for great concern.
- As our population increases, so do:
  - our demand for resources
  - our waste production
  - our environmental damage

8.3 The Future of World Population: Some Projections and Concerns

- Predicting the future size of the world's population is difficult.
- It is likely that the population of the world will increase dramatically before it stabilizes.
- This increase could bring about massive changes in the environment.

CH08-20

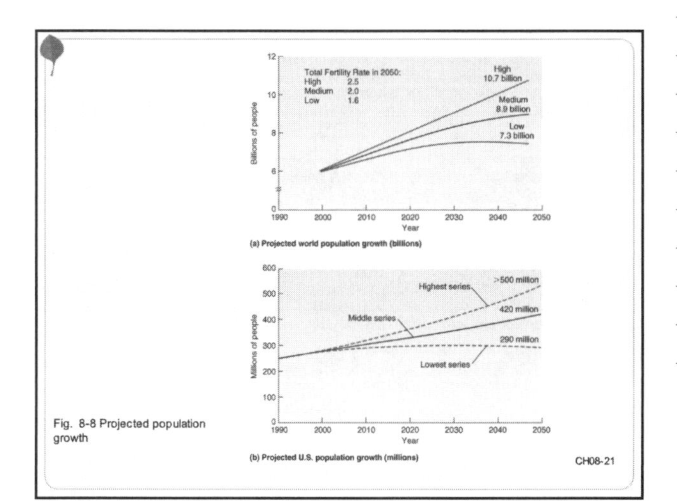

Fig. 8-8 Projected population growth

CH08-21

Notes

Notes

❖ Population Growth in the Less Developed World: Why Should We Worry?

- Population growth in the less developed nations of the world has many social, economic, and environmental impacts—serious impacts that affect them as well as us.

CH08-22

❖ A World of Possibilities

- Most experts agree that the human population cannot grow indefinitely.
- Some countries will very likely make a smooth transition to a stable population size.
- Others may experience periodic crashes that will eliminate large numbers of people.
- Some countries may overshoot the carrying capacity and destroy their ability to support people so drastically that their populations may fall to much lower levels.

CH08-23

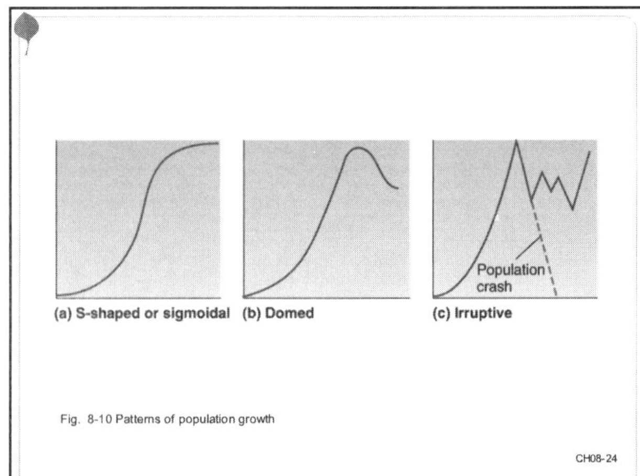

Fig. 8-10 Patterns of population growth

CH08-24

Chapter 9: Stabilizing the Human Population: Strategies for Sustainability

Notes

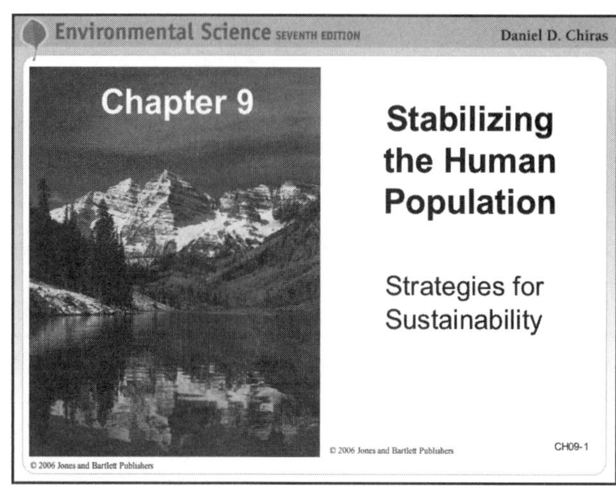

9.1 Achieving a Sustainable Human Population: The Challenges

- Most people consider the main challenge of achieving a sustainable human population to be finding acceptable means of reducing the population growth rate to stabilize the world population.
- Many others, however, argue that to live sustainably on the Earth, we must reduce human numbers through humane, socially acceptable means.

CH09-2

9.2 Stabilizing the Human Population: Some Strategies

- Stabilizing the human population will require a number of measures besides access to contraception and family planning that attack the root causes of rampant population growth.

CH09-3

Notes

❖ Economic Development and the Demographic Transition

- Economic development can be a powerful force for reducing population growth.
- Economic development caused a shift in population growth in the more developed nations.
- This change took many decades and substantial resources, which the less developed nations of the world do not have.

CH09-4

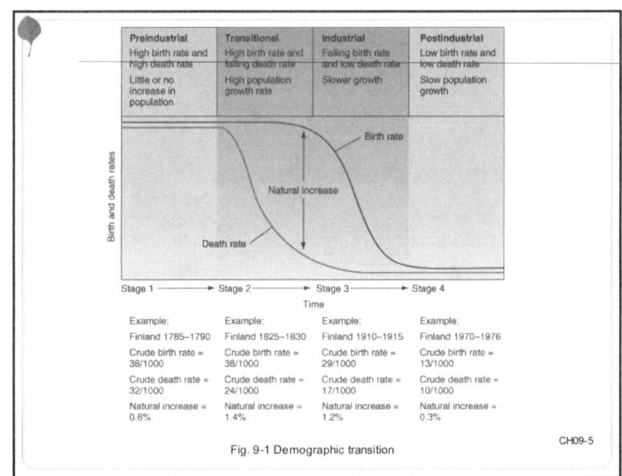

Fig. 9-1 Demographic transition

CH09-5

❖ Family Planning and Population Stabilization

- Family planning measures permit couples to determine the number and spacing of children to determine family size.
- These measures are vital to global efforts to reach a sustainable human population.
- Vital components of a global strategy to reduce fertility and population growth include:
  - small-scale sustainable economic development
  - jobs for women
  - efforts to promote equality
  - improvements in health care for women

CH09-6

❖ Sustainable Populations in the More Developed Countries

- Most of the attention on curbing population growth is focused on the less developed nations—the largest sector of the global population.
- The more developed nations have an important role because of their high level of per capita consumption and environmental impact.

CH09-7

Notes

❖ Sustainable Populations in the More Developed Countries (cont.)

- The impact of a population depends on many factors, most importantly:
  - the size of the population
  - per capita consumption (how much citizens consume on average)
  - the resources used and pollution produced to meet needs

CH09-8

❖ Sustainable Populations in the More Developed Countries (cont.)

- The industrial nations can do many things to help build a sustainable future, for example:
  - reducing their own growth, use of resources, and pollution
  - Assisting less developed countries (LDCs) through financial aid, especially for family planning and sustainable development
  - sharing information and technology

CH09-9

Notes

❖ Creating Sustainable Populations in the Less Developed Countries

- Numerous private and governmental organizations spend millions of dollars a year to support programs to slow the population growth in LDCs through:
  - family planning
  - sustainable economic development
- The **International Planned Parenthood Federation** and the **United Nations Fund for Population Activities** are two examples of such organizations.

CH09-10

9.3 Overcoming Barriers

- Three primary barriers lie in the way of achieving a sustainable human population:
  - Psychological and cultural
  - Educational
  - Religious

CH09-11

❖ Psychological Barriers

- In LDCs, children are often seen as an asset to their parents.
  - Childbearing enhances a woman's social status.
- Having many children seems desirable because mortality tends to be higher, too.
- In more developed nations, children are valued, but are viewed as a bit of an economic drain, and a woman's status is not so heavily dependent on childbearing.

CH09-12

❖ Educational Barriers

- As a general rule, the higher the level of education in a population, the lower its fertility rate.
- Education and careers decrease the number of childbearing years and open up many options besides childbearing.

CH09-13

❖ Religious Barriers

- Religions influence population both positively and negatively.
- Some religions openly denounce efforts to control population growth; others openly support smaller families. Some have no official view.

CH09-14

9.4 Ethics and Population Stabilization

- Many people think that the right to have children is a fundamental personal freedom.
- Others believe that such individual rights are superseded by the collective rights of present and future generations to:
  - a clean, healthy environment
  - adequate food, shelter, and clothing

CH09-15

Notes

Notes

9.5 Status Report: Progress and Setbacks

- ↑ World population growth rates have declined since the early 1960s—from 2% down to 1.2% in 2005.
- ↑ Two international conferences set forth strategies to improve the status of women.
- ↓ Population growth has resulted in overcrowded urban areas.
- ↓ 1/3 of growth rate decline is the result of AIDS.
- ↓ 1/3 of the World's population is under 15 and about to enter its reproductive years.
- ↓ Population growth is still rapid in LDCs without the resources to handle the growth.

CH09-16

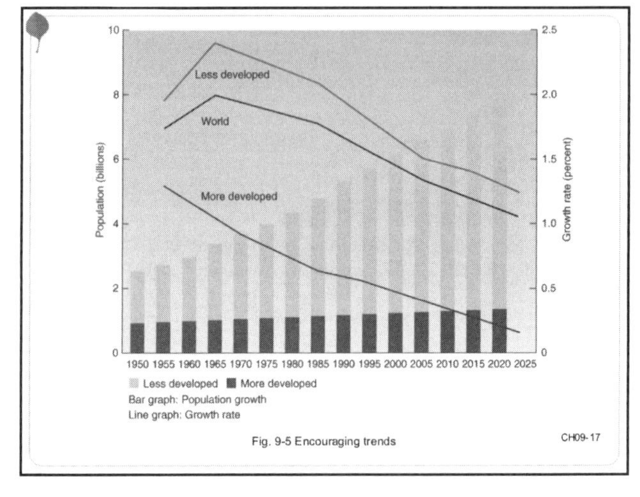

Fig. 9-5 Encouraging trends

CH09-17

Chapter 10: Creating a Sustainable System of Agriculture to Feed the World's People

Notes

Environmental Science SEVENTH EDITION   Daniel D. Chiras

Chapter 10

Creating a Stable System of Agriculture to Feed the World's People

CH10-1

10.1 Hunger, Malnutrition, Food Supplies, and the Environment

- A large segment of the world's people (most of whom live in Asia, Africa, and Latin America):
  - either do not get enough to eat
  - or fail to get all of the nutrients and vitamins they need
  - or both
- Nutritional deficiencies make people more susceptible to infectious disease and, if they are severe enough, can cause death.

CH10-2

❖ Hunger, Poverty, and Environmental Decay

- Hunger and malnutrition cause mental and physical retardation.
- This may contribute to widespread poverty, which in turn contributes to environmental destruction.

Fig. 10-1 Kwashiorkor

CH10-3

Notes

❖ Declining Food Supplies

- Grain production per capita has been on the decline for over a decade and a half.
- This trend bodes poorly for those suffering from hunger and malnutrition—as well as for those trying to provide food for the ever-growing human population.

CH10-4

❖ The Challenge Facing World Agriculture: Feeding People/Protecting the Planet

- The challenge today is to find sustainable ways to feed current and future world residents

CH10-5

10.2 Understanding Soils

❖ What Is Soil?

- Soils consist of four components:
  - inorganic materials
  - organic matter
  - air
  - water

CH10-6

❖ How Is Soil Formed?

- Soil formation is a complex process involving an interaction among:
  - Climate
  - The parent material, which contributes the mineral components of soil
  - Biological organisms

CH10-7

❖ The Soil Profile

- Soils are typically arranged in layers.
- For agriculture, the most important are the upper two layers:
  - the O horizon, which accumulates organic waste from plants and animals
  - the A horizon, the topsoil.

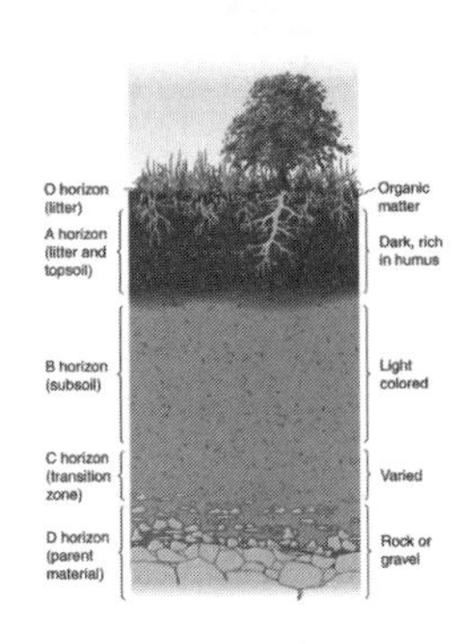

Fig. 10-3 Soil profile

CH10-8

10.3 Barriers to a Sustainable Agricultural System

❖ Soil Erosion

- Soil is vital to the success of a nation, indeed the world.
- Agricultural soils are being lost at record rates in many countries – a trend that is clearly unsustainable.

Fig. 10-4 soil erosion

CH10-9

Notes

Notes

❖ Desertification: Turning Cropland into Desert

- Throughout the world, cropland, rangeland, and pasture are becoming too dry to use because of:
  - climate change (natural and human-induced)
  - poor land management practices such as overgrazing

CH10-10

❖ Desertification: Turning Cropland into Desert (cont).

- Desertification and soil erosion are destroying agricultural land worldwide.
- This contributes to food shortages and reduces our ability to meet future demands for an expanding human population.

CH10-11

❖ Farmland Conversion

- Each year, millions of hectares of productive agricultural land are lost to human development worldwide.
- This phenomenon is called farmland conversion.

CH10-12

❖ Declines in Irrigated Cropland Per Capita

- Irrigated cropland supplies enormous amounts of food to the world's people.
- The amount of irrigated cropland per capita is on the decline.
- This trend bodes poorly for world food production.
- Measures that increase the efficiency of water use may prove helpful in providing an adequate supply of irrigation water.

CH10-13

❖ Waterlogging and Salinization

- Irrigation can cause waterlogging, the buildup of excess water in the soil.
- Waterlogging suffocates plants.
- It may also cause salinization, the deposition of salts that are toxic to most plants.
- Waterlogging and salinization affect many millions of hectares of land worldwide.

CH10-14

❖ Declining Genetic Diversity in Crops and Livestock

- The number of species of cultivated plants and domestic animals has declined dramatically.
- Reducing diversity results in huge monocultures of genetically similar plants, which make crops more susceptible to:
  - Disease
  - Adverse weather
  - Insects and other pests
- It makes plants more dependent on chemical pesticides.

CH10-15

Notes

Notes

❖ Declining Genetic Diversity in Crops and Livestock

- The Green Revolution was a worldwide effort to improve the productivity of important food crops: **wheat and rice**.
- It succeeded in its primary objectives but created a steady decline in genetic diversity, which makes world food production more vulnerable to disruption.

CH10-16

❖ Declining Genetic Diversity in Crops and Livestock

- We are losing many wild plant species that gave rise to modern crop species throughout the world, especially in the tropics.
- This erodes our capacity to improve crops and make them more resistant to pests, disease, and drought.

CH10-17

❖ Politics, Agriculture, and Sustainability

- The problems facing world agriculture are not all technical.
- Some result from inadequate or self-defeating policies and governmental intervention.
- Lawmakers throughout the world have unwittingly facilitated the creation of an unsustainable system of agriculture.

CH10-18

10.4 Solutions: Building a Sustainable Agricultural System

- A sustainable system of agriculture consists of practices that produce high-quality food in ways that protect the long-term health and productivity of soils.
- Creating such a system will require a multifaceted approach, including measures to slow and perhaps stop the growth of the human population.

CH10-19

❖ Protecting Existing Soil and Water Resources (cont.)

- Protecting soil and water resources is the first line of defense in meeting present and future needs for food.
- One of the highest priorities in making the transition to a sustainable system of agriculture is putting an end to excessive soil erosion.
- Fortunately, there are many simple yet effective measures that can ensure a sustainable erosion rate.

CH10-20

❖ Protecting Existing Soil and Water Resources (cont.)

- Reducing the amount of land disturbance by minimizing tillage protects the soil from the erosive forces of wind and rain.
- This technique, while effective in reducing energy demand and erosion, often requires additional chemical herbicides to control weeds.

Fig. 10-9 Minimum tillage planter

CH10-21

Notes

Notes

❖ Protecting Existing Soil and Water Resources (cont.)

- Planting crops perpendicular to the slope—that is, along the land **contour** lines—reduces soil erosion and increases water retention.
- Crops can be planted in alternating strips, a practice called **strip cropping**.
- When strip cropping is combined with contour farming, this technique greatly reduces soil erosion.

Fig.10.12 Strip cropping (alfalfa and corn) protects against soil erosion.

CH10-22

❖ Protecting Existing Soil and Water Resources (cont.)

- Terraces, small earthen embankments that run across the slope of the land, greatly reduce soil erosion.
- Gullies form quickly on hilly terrain and grow rapidly.
- If gullies form, they can be regraded and replanted with fast-growing species to prevent their expansion.

Fig. 10.13 Terracing in Asia

CH10-23

❖ Protecting Existing Soil and Water Resources (cont.)

- **Shelterbelts** are rows of trees planted along the perimeter of fields to block wind and reduce soil erosion.
- Shelterbelts have the added benefit of preventing snow from blowing away from fields, thus increasing soil moisture content.
- Shelterbelts provide habitat for useful species, such as insect-eating birds.

Fig.10.14 Shelterbelts

CH10-24

❖ Protecting Existing Soil and Water Resources (cont.)

- Farmers are sometimes reluctant to take measures to control erosion because of their costs.
- Carefully crafted government policies can provide economic incentives to protect soil erosion.
- Many measures that protect soil from erosion also make it less susceptible to desertification.

CH10-25

❖ Protecting Existing Soil and Water Resources (cont.)

- When combined with measures to reduce global warming, these steps could help to slow desertification.
- Numerous techniques are available to prevent farmland conversion.
- Water efficiency measures help free up water to expand irrigated cropland.
- More frugal application of irrigation water to crops and special drainage systems can reduce salinization and waterlogging.

Fig. 10-15 Trickle irrigation system

CH10-26

❖ Soil Enrichment Programs

- Farming mines the soil, robbing it of valuable nutrients.
- Numerous methods such as applying organic fertilizer and rotating crops can replenish nutrients and maintain the health of the soil.
- Use of organic fertilizers helps farmers maintain or even improve soil conditions and boost crop production.
- This strategy also returns nutrients to the soil.

CH10-27

Notes

Notes

❖ Soil Enrichment Programs (cont.)

- Synthetic fertilizers help boost soil fertility.
- But they only partially replenish agricultural soils, because they contain only three of many nutrients needed for healthy soil:
  - Nitrogen
  - Phosphorus
  - Potassium

CH10-28

❖ Soil Enrichment Programs (cont.)

- Crop rotation – alternating crops planted in a field one season after another – offers many benefits.
- Planting the proper crops can help replenish soil nutrients.
- It also helps reduce erosion, pest damage, and the need for costly and potentially harmful pesticides.

CH10-29

❖ Increasing the Amount of Land in Production

- Grasslands and forests can be converted to farmland to meet the rising demand for food.
- In many parts of the world, especially in the more developed nations, farmland reserves are small.
- Even in countries where there is an abundance of reserve land, much of this land is covered with poor soils.
- Furthermore, the ecological cost of converting wild land to farmland would be **enormous**.

CH10-30

❖ Increasing the Productivity of Existing Land: Developing Higher-Yield Plants and Animals

- Numerous efforts are under way to increase the yield of plants and the growth rate of animals to increase food production.
- Geneticists can improve plant and animal strains by selective breeding and genetic engineering.
- Selective breeding has been used for hundreds of years.
- Genetic engineering is the deliberate transfer of genes from one organism to another.

CH10-31

❖ Increasing the Productivity of Existing Land: Developing Higher-Yield Plants and Animals (cont.)

- Genetic engineering has an enormous potential but poses many ethical questions and may create some serious environmental problems.
- Protecting wild plant species through habitat protection and special seed banks is essential to the future of agriculture.
- It helps to preserve genes that can improve crop yields by providing resistance to insects, disease, and drought.

CH10-32

❖ Developing Alternative Foods

- Many native plant and animal species could be used to provide food.
- Native animals offer many benefits over domestic livestock, including their resistance to disease-causing organisms.

CH10-33

Notes

Notes

❖ Developing Alternative Foods (cont.)

- Most of the world's commercially important saltwater fish populations are in decline and in danger of being seriously depleted.
- The decline in wild fish populations has forced many countries to grow fish commercially in ponds, lagoons, and other water bodies.

Table 10-1

Top 10 Species by Weight, 1970, 1980, 1992, and 2000 (catch in million tons)

1970		1980		1992		2000	
1. Peruvian anchovy	13.1	1. Alaska pollock	4.0	1. Peruvian anchovy	5.5	1. Peruvian anchovy	9.7
2. Atlantic cod	3.1	2. South American pilchard	3.3	2. Alaskan pollock	5.0	2. Alaskan pollock	2.7
3. Alaska pollock	3.1	3. Chub mackerel	2.7	3. Chilean jack mackerel	3.4	3. Skipjack tuna	2.0
4. Atlantic herring	2.3	4. Japanese pilchard	2.6	4. South American pilchard	3.1	4. Capelin	1.96
5. Chub mackerel	2.0	5. Capelin	2.6	5. Japanese pilchard	2.5	5. Atlantic herring	1.87
6. Capelin	1.5	6. Atlantic cod	2.2	6. Capelin	2.1	6. Japanese anchovy	1.85
7. Haddock	0.9	7. Chilean jack mackerel	1.3	7. Silver carp[1]	1.6	7. Chilean jack mackerel	1.75
8. Cape hake	0.8	8. Blue whiting	1.1	8. Atlantic herring	1.5	8. Blue whiting	1.6
9. Atlantic mackerel	0.7	9. European pilchard	0.9	9. Skipjack tuna	1.4	9. Chub mackerel	1.47
10. Saithe	0.6	10. Atlantic herring	0.9	10. Grass carp[2]	1.3	10. Largehead hairtail	1.45

Source: U.N. Food and Agriculture Organization.

[1]Raised on freshwater fish farms; all others are wild marine species.

[2]Raised on freshwater fish farms; all others are wild marine species.

CH10-34

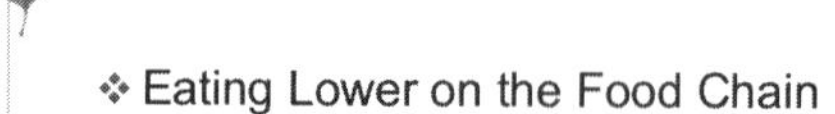

❖ Eating Lower on the Food Chain

- Efforts to feed the world's people should focus on food sources low on the food chain—plants and plant products.
- Far more people can be fed on a vegetarian diet than on a meat-based one.

CH10-35

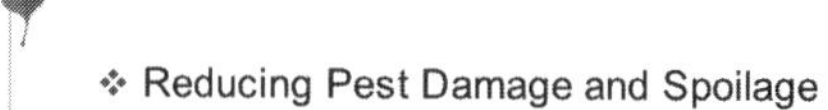

❖ Reducing Pest Damage and Spoilage

- Much of the world's food production is consumed by pests or rots in storage or in transit.
- Improvements in transit and storage, such as refrigeration, can greatly boost food supplies.

CH10-36

❖ Creating Agricultural Self-Sufficiency in Less Developed Nations

- Many LDCs have lost their ability to produce food as a result of:
  - overpopulation
  - farmland deterioration
  - economic and trade policies
- Reversing these trends could help nations become more self-reliant, which is vital for building a more sustainable future.

CH10-37

❖ Legislation and New Policies: Political and Economic Solutions

❖ Ending War

- Violent conflicts among peoples can greatly disrupt the production and distribution of food, often long after war has ended.

CH10-38

Notes

Notes

11.1 Biodiversity: Signs of Decline

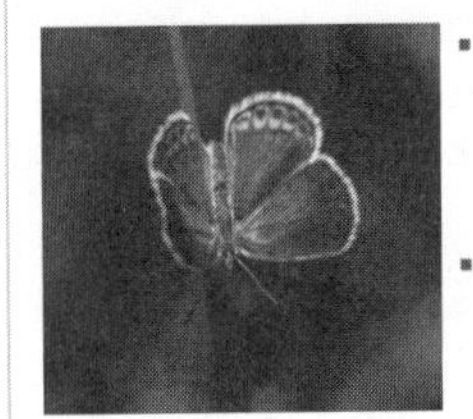
Karner Blue Butterfly (Courtesy of the USFWS)

- Many species of plants and animals face extinction today as a result of human activities.
- Although extinction has occurred since the dawn of time, modern extinctions are occurring at a rate much faster than is biologically sustainable.

CH11-2

11.2 Causes of Extinction and the Decline in Biodiversity

- Many factors contribute to the loss of species, but the two most important are:
  - the destruction and alteration of habitat
  - commercial harvesting

Fig. 11.2 Habitat fragmentation

CH11-3

Notes

❖ Physical Alteration of Habitat

- Virtually all human activities alter the environment, changing the biotic and abiotic conditions and fragmenting habitat.
- Habitat alteration is the #1 cause of species extinction.
- The most dramatic changes occur in biologically rich areas:
  - Tropical rain forests
  - Wetlands
  - Estuaries
  - Coral reefs

CH11-4

❖ Commercial Hunting and Harvesting

- Commercial hunting and harvesting of wild species have occurred for centuries.
- They represent the second largest threat to the world's animal species.
- This includes past activities, such as whale hunting, and present activities, such as commercial fish harvesting and poaching of endangered species.

CH11-5

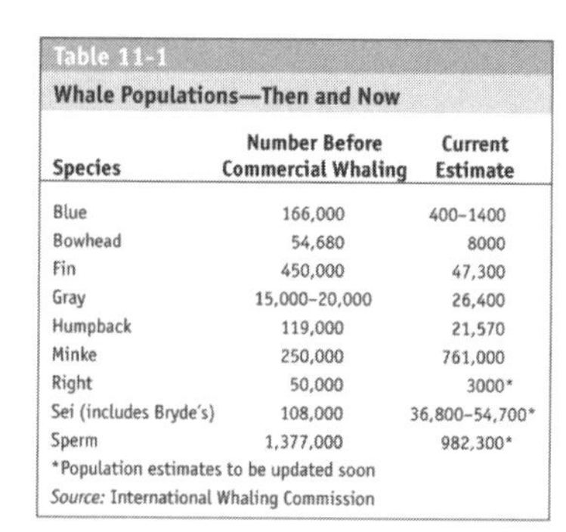

Table 11-1
Whale Populations—Then and Now

Species	Number Before Commercial Whaling	Current Estimate
Blue	166,000	400–1400
Bowhead	54,680	8000
Fin	450,000	47,300
Gray	15,000–20,000	26,400
Humpback	119,000	21,570
Minke	250,000	761,000
Right	50,000	3000*
Sei (includes Bryde's)	108,000	36,800–54,700*
Sperm	1,377,000	982,300*

*Population estimates to be updated soon
Source: International Whaling Commission

CH11-6

❖ The Introduction of Foreign Species

- Plant and animal species introduced into new regions may thrive because of the favorable conditions and low environmental resistance.
- They often out-compete and eliminate native species.
- Islands are especially vulnerable to foreign species.

CH11-7

Notes

❖ Pest and Predator Control

- Chemical pesticides, sprayed on farms and other areas to control insect pests, and predator control programs have had a profound impact on native species.

CH11-8

❖ The Collection of Animals and Plants for Human Enjoyment, Research, and Other Purposes

- Millions of plants and animals are taken from the wild and imported into developed countries for
  - zoos
  - private collections
  - pet shops
  - research
- This contributes to the worldwide loss of species.

CH11-9

Notes

❖ Pollution

- Pollution alters the physical and chemical nature of the environment in ways that impair the survival of many species.
- Pollution and climate change (caused by pollution) may be altering the health of the world's coral reefs.
  - This may cause widespread decimation if something is not done to reverse the trend.

CH11-10

❖ Biological Factors That Contribute to Extinction

- Many biological characteristics of organisms determine how vulnerable they are to human impacts on the environment, such as:
  - the number of offspring they produce
  - the size of their range
  - their tolerance for people
  - their degree of specialization

CH11-11

❖ The Loss of Keystone Species

- Keystone species are organisms upon which many other species in an ecosystem depend.
- The loss of a single keystone species may have a devastating effect on other organisms.

Sea otter

CH11-12

Notes

❖ A Multiplicity of Factors

- Many factors acting together contribute to the loss of biodiversity.
- These factors may synergize to produce a level of devastation far greater than anticipated.

CH11-13

11.3 Why Protect Biodiversity?

- Arguments for protecting endangered species and preserving biodiversity can be made on both utilitarian and non-utilitarian grounds.

Fig. 11.10a The snail darter.

CH11-14

❖ Aesthetics and Economics

- Some people believe that we should save other species because they are a source of beauty and pleasure.
- In addition, this can provide an economic benefit through activities such as ecotourism and bird watching.

CH11-15

Notes

❖ Food, Pharmaceuticals, Scientific Information, and Products

- Wild plants and animals are a valuable economic resource.
- They could provide:
  - new food sources to feed the growing human population
  - genes that could improve crop species
  - new medicines to combat disease
  - scientific knowledge
  - an assortment of products useful to humans

CH11-16

❖ Protecting Free Services and Saving Money

- Protecting natural systems helps preserve many ecological services such as flood control and water pollution abatement.
- These services are very costly to replace with engineered solutions.

Fig. 11-9 Mangrove swamp

CH11-17

❖ Ethics—Doing the Right Thing

- To many people, preservation of other species is ethically appropriate.
- Protecting them honors their right to exist and is therefore ethically correct.

Piping Plover

CH11-18

11.4 How to Save Endangered Species and Protect Biodiversity— A Sustainable Approach

- Protecting endangered species and preserving the world's dwindling biodiversity will require many actions ranging from **short-term protective** measures to **long-term preventive** efforts.

CH11-19

❖ A Question of Priorities: Which Species Should We Protect?

- Although many species are endangered, most resources are expended on the most appealing or most visible ones.
- Many ecologically important species could vanish if efforts are not broadened.

CH11-20

❖ Stopgap Measures: First Aid for an Ailing Planet

- The U.S. Endangered Species Act is a model of species protection legislation.
- However, it is essentially an emergency measure aimed at saving species already endangered or threatened with extinction.

ENDANGERED SPECIES ACT OF 1973[1]

AN ACT To provide for the conservation of endangered and threatened species of fish, wildlife, and plants, and for other purposes.

Be it enacted by the Senate and House of Representatives of the United States of America in Congress assembled, That this Act may be cited as the "Endangered Species Act of 1973".

TABLE OF CONTENTS

Sec. 2. Findings, purposes, and policy.
Sec. 3. Definitions.
Sec. 4. Determination of endangered species and threatened species.
Sec. 5. Land acquisition.
Sec. 6. Cooperation with the States.
Sec. 7. Interagency cooperation.
Sec. 8. International cooperation.
Sec. 8A. Convention implementation.
Sec. 9. Prohibited acts.
Sec. 10. Exceptions.
Sec. 11. Penalties and enforcement.
Sec. 12. Endangered plants.
Sec. 13. Conforming amendments.
Sec. 14. Repealer.
Sec. 15. Authorization of appropriations.
Sec. 16. Effective date.
Sec. 17. Marine Mammal Protection Act of 1972.
[Sec. 18. Annual cost analysis by the Fish and Wildlife Service.[2]]

FINDINGS, PURPOSES, AND POLICY

SEC. 2. (a) FINDINGS.—The Congress finds and declares that—
(1) various species of fish, wildlife, and plants in the United States have been rendered extinct as a consequence of economic growth and development untempered by adequate concern and conservation;
(2) other species of fish, wildlife, and plants have been so depleted in numbers that they are in danger of or threatened with extinction;
(3) these species of fish, wildlife, and plants are of esthetic, ecological, educational, historical, recreational, and scientific value to the Nation and its people;
(4) the United States has pledged itself as a sovereign state in the international community to conserve to the extent

[1] As amended by P.L. 94-325, June 30, 1976; P.L. 94-359, July 12, 1976; P.L. 95-212, December 19, 1977; P.L. 95-632, November 10, 1978; P.L. 96-159, December 28, 1979; 97-304, October 13, 1982; P.L. 98-327, June 25, 1984; and P.L. 100-478, October 7, 1988; P.L. 100-653, November 14, 1988; and P.L. 100-707, November 23, 1988.
[2] Bracketed material does not appear in Act. Sec. 1012 of P.L. 100-478, 102 Stat. 2314, October 7, 1988, added sec. 18 of the Act but did not conform the table of contents of the Act.

1

CH11-21

Notes

Notes

❖ Stopgap Measures: First Aid for an Ailing Planet (cont.)

- Zoos are an important player in a global effort to protect endangered species.
- They not only house many endangered species, protecting them from extinction, they are breeding many species for eventual release into protected habitat.

CH11-22

❖ Long-Term Preventive Measures

- Many stopgap measures are required to save species from immediate extinction.
- In the long run, preserving biodiversity requires preventive actions, including steps to help **restructure human systems** for sustainability.

CH11-23

❖ Long-Term Preventive Measures (cont.)

- Protecting biodiversity will be best achieved by efforts that address the root causes of the crisis of unsustainability:
  - our inefficient use of resources
  - continued population growth
  - reliance on fossil fuels
  - failure to recycle extensively
  - our lack of attention to restoration
- Addressing these issues will protect plants and animals and bring many other benefits to society.

CH11-24

❖ Long-Term Preventive Measures (cont.)

- Setting aside high-biodiversity areas for permanent protection will help to protect species from extinction and will help preserve biodiversity.
- Unfortunately, the majority of the most biologically diverse areas are located in the less developed nations, which lack the financial resources to protect them.

CH11-25

Notes

❖ Long-Term Preventive Measures (cont.)

- Islands of habitat are vital to protect species, but they may not be enough to prevent species loss.
- Buffer zones between human activities and protected areas may provide an additional measure of protection.
- Wildlife corridors are also proving vital in the efforts to protect species diversity.

CH11-26

❖ Long-Term Preventive Measures (cont.)

- Protected lands can be sustainably harvested by indigenous peoples, a strategy that protects biodiversity.
- Saving species and protecting biodiversity will require many improvements in wildlife management.
  - Especially the adoption of ecosystem management, which takes a broader view of species protection.

CH11-27

Notes

❖ Personal Solutions

- Saving species and protecting biodiversity require personal actions.
- We cannot wait for government or business to solve the problems for us.

CH11-28

Chapter 12: Grasslands, Forests, and Wilderness: Sustainable Management Strategies

Notes

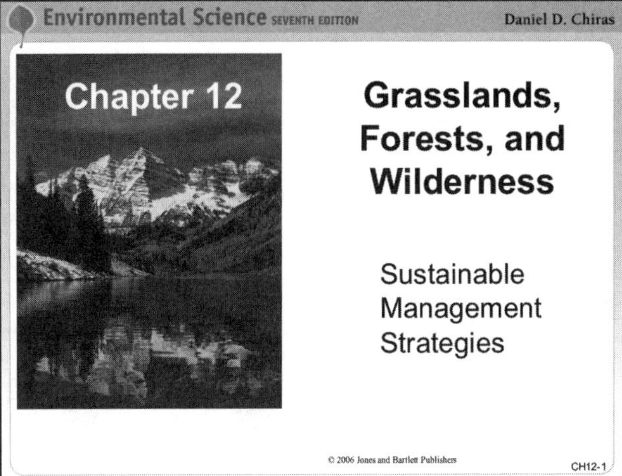

12.1 The Tragedy of the Commons

- Communal resources – resources held in common by people – often deteriorate as individuals become caught up in a cycle of self-gratification. For example:
  - Land
  - Air
  - Oceans
- Personal gain dictates actions that have negative effects shared by all who use communal property.
- Privately owned lands also deteriorate as a result of ignorance, greed, and other factors.

CH12-2

12.2 Rangelands and Range Management: Protecting the World's Grasslands

- Rangelands – grasslands on which livestock graze – are an important element of the global agricultural system.
- When properly managed, they can be a sustainable food source.

CH12-3

Notes

❖ An Introduction to Rangeland Ecology

- Grasses form the base of the food chain on rangelands.
- These hardy species are adapted to periodic drought, fire, and grazing as long as care is taken to protect the metabolic reserve of the plant.

Ungrazed → Grazed → Recovery

Metabolic reserve; Metabolic reserve intact; Metabolic reserve

(a)

Ungrazed → Overgrazed → Death

Metabolic reserve; Most of metabolic reserve eaten; Death

(b)

Fig. 12-3 The anatomy of grass

CH12-4

❖ The Condition of the World's Rangeland

- A large percentage of publicly and privately owned rangeland in the U.S. and other countries has been degraded because of unsustainable land management practices such as **overgrazing**.

CH12-5

❖ Rangeland Management: A Sustainable Approach

- Grasses are well adapted to grazing pressure.
- Grasslands and herbivores can coexist in a sustainable relationship that is beneficial to both.
- Rangeland and pasture use must be adjusted according to the carrying capacity of the land, which varies with the weather from one year to the next.
- Those who cannot adjust their herd size run the risk of lowering the carrying capacity of their land and even destroying grazing opportunities.

CH12-6

❖ Rangeland Management: A Sustainable Approach (cont.)

- Cattle can be shifted from one pasture to another to permit grasses to mature and produce seeds.
- This method enhances the condition of rangeland and may increase the carrying capacity in the long run.
- Fencing and careful distribution of water sources and salt licks can help promote a more uniform use of rangeland and protect some areas from serious degradation.

CH12-7

❖ Rangeland Management: A Sustainable Approach (cont.)

- Restoration of degraded grasslands is an essential element of building a sustainable system of livestock production.
- Efforts to boost the productivity of land, including periodic burns, also help.

A a prairie burn in Kansas

CH12-8

❖ Revamping Government Policies

- Many ill-conceived government policies result in the deterioration of publicly owned rangeland.
- To promote sustainable use of grasslands, government policies should be based on objective scientific criteria.

❖ Sustainable Livestock Production

- In many countries, livestock are raised in pens and fed grains (grown on land) that could be used to feed large numbers of people.

CH12-9

Notes

Notes

12.3 Forests and Forest Management

- The world's forests provide many social, economic, and environmental benefits.
- A large portion of the world's forests have been logged or disturbed.
- Very little forested land is under permanent protection.

CH12-10

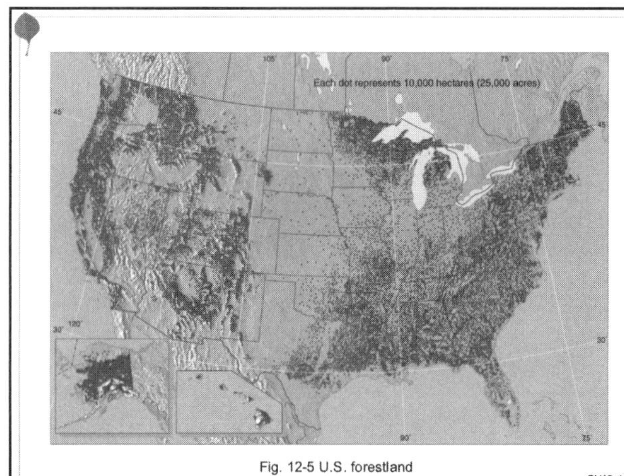

Fig. 12-5 U.S. forestland

CH12-11

❖ Status Report on the World's Forests

- About one-half of the world's forests have been cut.
- The land they once occupied has already been converted to other uses, mostly farming, or undergone severe deterioration.
- Deforestation continues today at a rapid pace and threatens the long-term sustainability of human civilization.
- Deforestation continues in:
  - tropical rain forests
  - northern coniferous forests
  - temperate deciduous forests

CH12-12

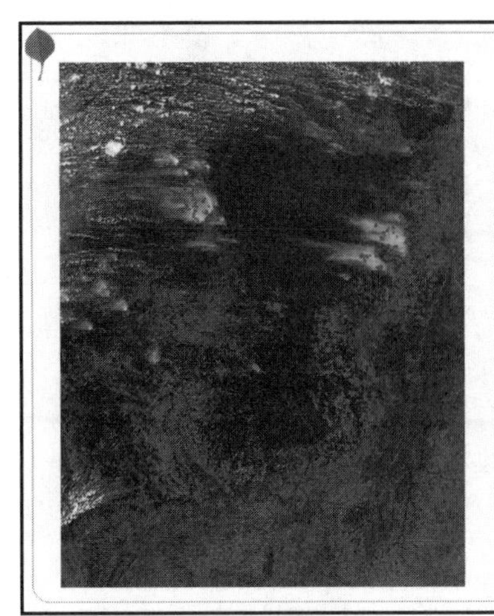

Deforestation: Dark green swaths of intact Amazon rainforest are nearly surrounded by agricultural land. The ragged lines show the progression of roads into the forest. The red outlines indicate fires. Some are likely routine burns of already-cleared farm and grazing land. The fires with thick plumes of smoke close to the new roads and at the edges of clearings are probably ongoing "slash and burn" deforestation.

(Image taken August 5, 2004, courtesy Jacques Descloitres, MODIS Rapid Response Team, NASA-Goddard Space Flight Center)

CH12-13

Notes

❖ Root Causes of Global Deforestation

- Deforestation results from many factors, including:
  - frontierism
  - a lack of knowledge of the importance of forests
  - population growth
  - poverty
  - inequitable land ownership

CH12-14

❖ Root Causes of Global Deforestation (cont.)

- Many nations still view forests as vast untapped reserves of wealth and actively promote their exploitation, in spite of the many ecological benefits.
- Ill-advised government policies, including below-cost timber sales, contribute to widespread deforestation and unsustainable forest management.
- These policies are often promoted by powerful economic interests that stand to gain from lenient timber-harvesting practices.

CH12-15

Notes

❖ An Introduction to Forest Harvesting and Management

- Trees are harvested primarily in four ways:
  - clear-cutting
  - strip-cutting
  - selective cutting
  - shelter-wood cutting

Fig. 12-6 Clear-cutting

CH12-16

❖ An Introduction to Forest Harvesting and Management (cont.)

- Clear-cutting removes entire forests quickly and efficiently.
- Some tree species such as pines, which grow in open sunny fields, are best harvested in clear-cuts.
- Clear-cuts benefit certain wildlife but tend to destroy and fragment the habitat of others.
- Clear-cutting creates ugly scars and can cause considerable environmental damage such as increased soil erosion.

CH12-17

❖ An Introduction to Forest Harvesting and Management (cont.)

- Clear-cutting can be carried out on a smaller scale to minimize visual and environmental impacts.
- One technique is known as strip-cutting—clear-cutting smaller, narrower strips of forest.

CH12-18

Notes

❖ An Introduction to Forest Harvesting and Management (cont.)

- Selective cutting takes place in multi-species forests with species whose seedlings grow best in shade.
- It reduces visual scarring but is expensive and time-consuming and can cause considerable damage to unharvested trees.

CH12-19

❖ An Introduction to Forest Harvesting and Management (cont.)

- Selective harvesting can be modified to correct its problems through shelter-wood cutting.
- This method, while more expensive, helps preserve multi-species forests.

CH12-20

❖ Creating a Sustainable System of Forestry

- Four measures are required to create a sustainable system of wood production:

1. reductions in demand for wood and wood products
2. sustainable management
3. establishment of forest preserves
4. restoration of forest land

CH12-21

Notes

❖ Creating a Sustainable System of Forestry (cont.)

- Demand for wood and wood products can be greatly reduced by:
  - controlling growth of the human population
  - using wood and wood products more efficiently
  - finding alternatives
  - recycling paper and wood materials

CH12-22

❖ Creating a Sustainable System of Forestry (cont.)

- Better management of existing forests, based on sound scientific principles, helps to create a more diverse and healthier forest that is less susceptible to disease and insects:
  - tree thinning
  - prescribed burns
  - replanting
- Certification programs can help promote sustainable forest management.

CH12-23

❖ Creating a Sustainable System of Forestry (cont.)

- Saving uncut or primary forests helps preserve biodiversity and protects nearby harvested forests from outbreaks of pests.
- Building a sustainable system of forestry will require efforts to replant millions of acres of forestland that has been cut and never replanted.

CH12-24

12.4 Wilderness and Wilderness Management

- Large tracts of wilderness, land largely untouched by humans, exist today.
- Pressure is mounting to develop many of these lands for timber, oil, and other resources.

Fig. 12-10 Denali National Park

CH12-25

❖ Why Save Wilderness?

- Wilderness offers many benefits to humans:
  - provides refuge from urban life
  - offers valuable ecological services
  - is home to many species of plants and animals
- Historically, wilderness has largely been viewed as either a source of resources or an impediment to human progress.
- These opposing views are at the root of the controversy over wilderness protection.

CH12-26

❖ Preservation: The Wilderness Act

- The U.S. has a long history of wilderness preservation that continues today through the Wilderness Act.
- This law directs federal agencies to establish wilderness areas and stipulates the type of human activities that are permitted on these lands.

President Theodore Roosevelt
Term of office: 1901-1909

CH12-27

Notes

Chapter 13: Water Resources: Preserving Our Liquid Assets and Protecting Aquatic Ecosystems

Notes

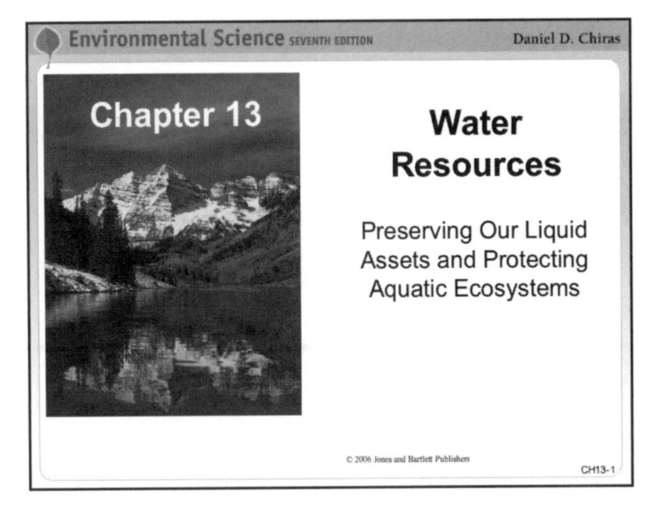

13.1 The Hydrological Cycle

- Water is a renewable resource, purified and distributed in the hydrological cycle, which is driven by solar energy.
- The Earth has an enormous quantity of water, but only a small fraction of it is available for human use.
- This fact underscores the importance of managing it wisely.

CH13-2

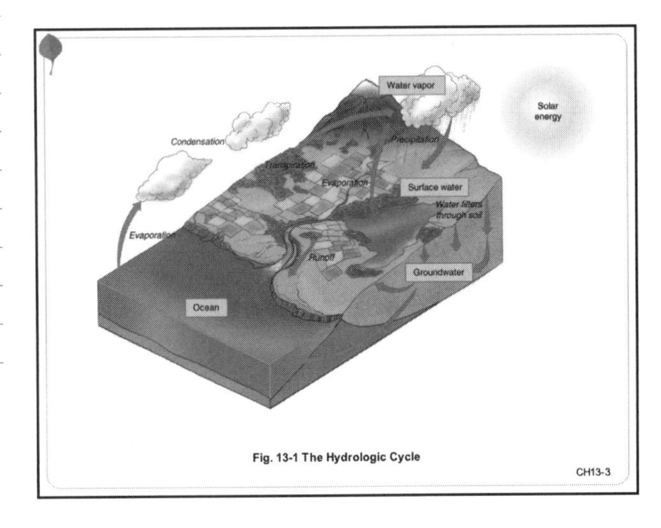

Fig. 13-1 The Hydrologic Cycle

CH13-3

Notes

13.2 Water Issues Related to Supply and Demand

❖ Where Does Water Come from and Who Uses It?

- Globally, agriculture and industry are the major users of water.
- Most water comes from surface water supplies like rivers, streams, and lakes.

CH13-4

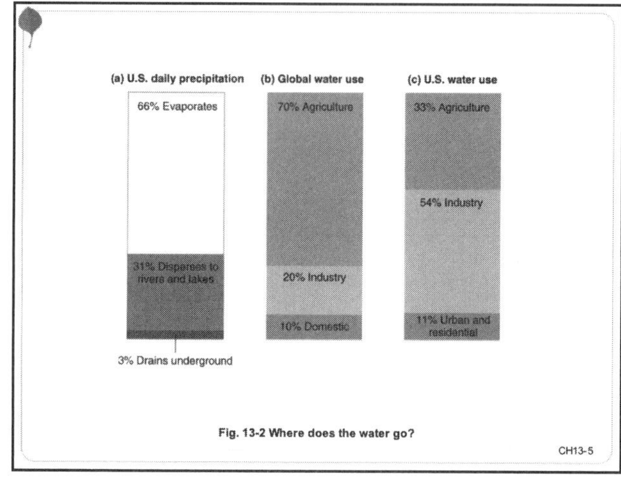

Fig. 13-2 Where does the water go?

CH13-5

❖ Water Shortages

- Water shortages occur virtually everywhere.
- They're most prevalent in areas that receive small amounts of precipitation, but can occur in any region in which demand exceeds existing water supplies.
- Factors likely to make water shortages even more prevalent in coming years:
  - population growth
  - agricultural expansion
  - demand for water by the industrial sector

CH13-6

❖ Drought and Water Shortages

- Droughts occur naturally, but may also result from human actions such as deforestation and overgrazing.
- Droughts reduce water supplies and create significant social, economic, and environmental problems.

CH13-7

❖ Impacts of the Water Supply System

- Water comes to us via an elaborate and costly system that has a tremendous impact on the environment.
- Understanding the system and its impacts may help us design more sustainable systems of water supply.

CH13-8

❖ Impacts of the Water Supply System (cont.)

- Much of the water removed from a source never returns to its source.

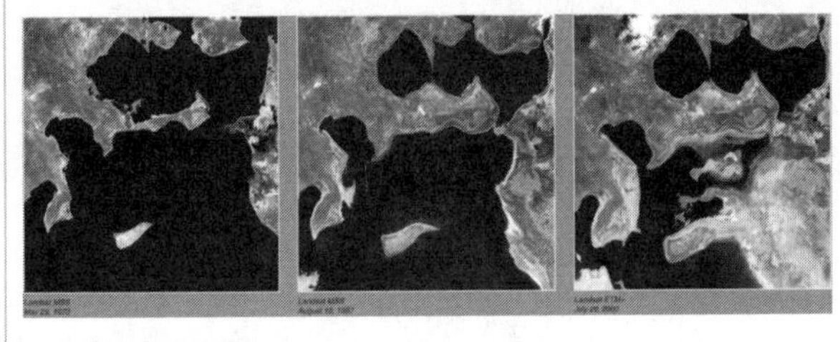

The Disappearing Aral Sea, Kazakhstan.
Image courtesy USGS Eros Data Center, based on data provided by the Landsat science team.

CH13-9

Notes

Notes

❖ Impacts of the Water Supply System (cont.)

- When demand becomes excessive, impacts start to become evident
  - Rivers run dry for extended periods
  - Habitat disappears
  - Population of aquatic species are reduced
  - Aquifers dry up
  - Salt water intrudes into freshwater aquifers

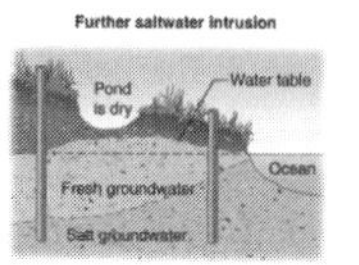

Fig. 13-4 Saltwater intrusion into groundwater

CH13-10

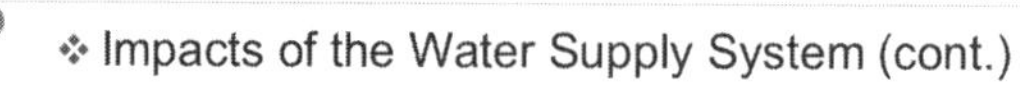

❖ Impacts of the Water Supply System (cont.)

- Dams and reservoirs have many benefits, such as flood control and water storage.
- They also have profound impacts on people, wildlife, and the economy.
- Ultimately, all dams have a finite life span because their reservoirs eventually fill with sediment.

Dworshak Dam, Idaho
Courtesy of the U.S. Army Corps of Engineers

CH13-11

❖ Creating a Sustainable Water Supply System

- To avert future water shortages, we need a sustainable water policy and management strategy based on four of the operating principles of sustainability:
  - conservation
  - recycling
  - restoration
  - population control
- Efforts to slow and perhaps stop the growth of the human population are vital to living within the planet's capacity to supply freshwater.

CH13-12

Notes

- Creating a Sustainable Water Supply System (cont.)
  - Water conservation is a relatively fast and highly cost-effective means of meeting demands for water.
  - Some of the largest gains can be made in agriculture and industry.
  - Domestic water conservation measures can also save enormous amounts of water.

CH13-13

- Creating a Sustainable Water Supply System (cont.)
  - Water can be purified and reused in industry, on farms, and in our homes.
  - Water recycling reduces pressure on surface water and groundwater supplies and reduces water pollution.
  - Wastewater can be used to recharge groundwater supplies or can be used to irrigate fields.

CH13-14

- Creating a Sustainable Water Supply System (cont.)
  - Restoring and protecting watersheds helps to prevent siltation in reservoirs and enhance groundwater recharge.

CH13-15

Notes

❖ Creating a Sustainable Water Supply System (cont.)

- Many government policies, especially subsidies, contribute to the unsustainable use of water.
- Changes in these policies can greatly increase the efficiency of water use and promote alternative water supply strategies, such as:
  - Conservation
  - Rain catchment
  - Recycling
  - Gray water systems
- Education is a key to creating individual responsibility and sustainable water-use patterns among citizens.

CH13-16

❖ Creating a Sustainable Water Supply System (cont.)

- As a last resort, rising demand can be met by developing new water supplies.
- New dams and diversion are one option, but their costs are enormous.
- Desalination plants are a likely candidate for coastal communities but are costly to operate.
- Excessive withdrawal of surface water can cause streams to dry up.
- Groundwater overdraft results in many problems.

CH13-17

13.3 Flooding: Problems and Solutions

❖ Flooding is a major problem in many areas of the world and appears to be on the rise, in part as a result of human activities.

New Orleans, LA

CH13-18

Causes of Flooding

- Floods result from natural events, for example, too much rain.
- They are also the result of changes in the environment caused by humans, notably:
  - changes to rainfall patterns
  - changes in the land surface that increase runoff

Midwest flood, June 1994

CH13-19

Fig. 13-1 Percolation-runoff ratio

CH13-20

Controlling Flooding

- Many solutions to flooding, like dams, levees, and streambed channelization, are unsustainable.
- They treat the symptoms, not the root causes of flooding.
- Preventive measures combined with restorative measures are far more effective in the long run.
- Such measures include:
  - controlling population growth
  - protecting watersheds
  - reducing global climate change

CH13-21

Notes

Notes

13.4 Wetlands, Estuaries, Coastlines, and Rivers

- Surface waters—lakes, rivers, and bays—are under assault.
- Their destruction affects available water supplies.
- It also destroys habitat for fish and other organisms that are important food sources for humans and other species.

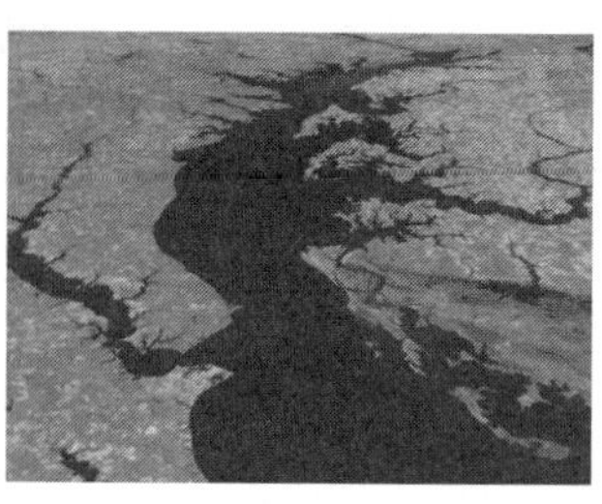

Fig. 13-14 Chesapeake Bay

CH13-22

- Wetlands
  - Wetlands are highly endangered ecosystems.
  - Wetlands are habitat for aquatic organisms, birds, and mammals.
  - They help control flooding, purify water, and buffer humans from storm surges.
  - Unfortunately, many people fail to understand their economic and ecological importance.

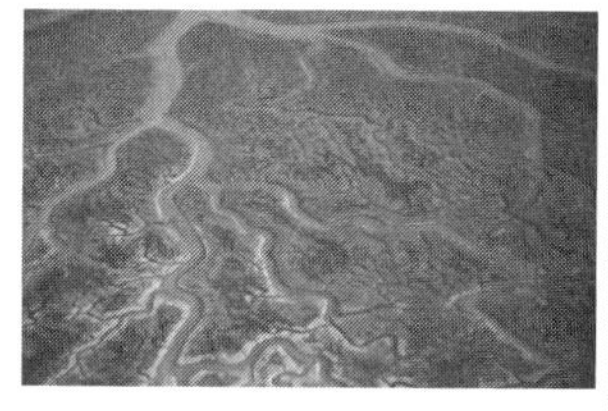

Humboldt Bay National Wildlife Refuge
Courtesy of the USFWS

CH13-23

- Wetlands, (cont.)
  - Wetlands are declining virtually everywhere, and massive areas have already been destroyed in the U.S. and abroad.
  - The rate of destruction in the U.S. has slowed considerably.
  - Many remaining wetlands are impaired because of:
    - pollution
    - invasion of exotic species
    - lack of adequate water flow

CH13-24

200,000
160,000
120,000
80,000
40,000
Hectares per year
1954–1974 1975–1985 1986–1997

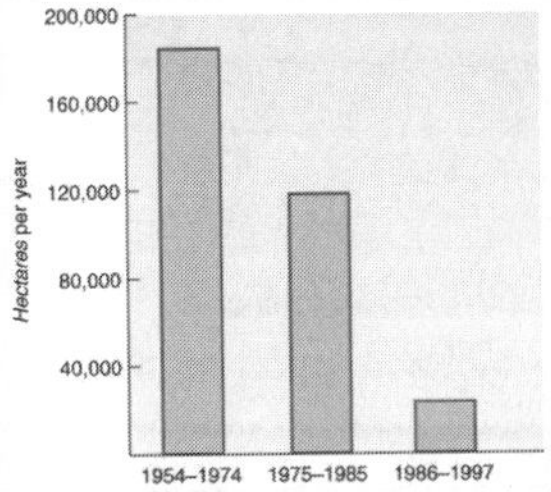

Fig. 13-17 Declining wetland losses

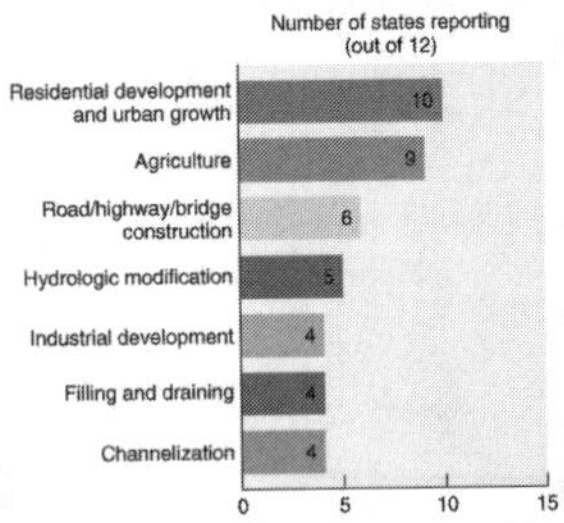

Number of states reporting (out of 12)

Source	Number of states reporting
Residential development and urban growth	10
Agriculture	9
Road/highway/bridge construction	6
Hydrologic modification	5
Industrial development	4
Filling and draining	4
Channelization	4

0 5 10 15

Fig. 13-18 Sources of wetland losses

CH13-25

❖ Wetlands, (cont.)

- Numerous efforts are under way at local, state, national, and even international levels to preserve and protect existing wetlands.
- Although these measures have not ended the destruction, they have greatly slowed it down.

CH13-26

❖ Estuaries

- Estuaries are the mouths of rivers and are biologically important life zones of great economic importance to humans.
- Because of its location downstream from many human activities, the estuarine zone is seriously endangered.
- Overharvesting of commercially important species, especially shellfish, in this zone carries enormous ecological and economic costs.

CH13-27

Notes

Notes

❖ Estuaries (cont.)

- Protecting the estuarine zone requires many efforts; these include:
  - establishing protected areas
  - carrying out numerous preventive measures such as:
    - pollution control
    - soil erosion controls on watersheds of rivers and their tributaries that empty into estuaries

CH13-28

❖ Barrier Islands

- Barrier islands are offshore islands made of sand that are popular sites for homes and other structures.
- The islands are very unstable and subject to violent storms that destroy homes and other buildings.
- U.S. government policy has promoted the development of barrier islands until quite recently.

CH13-29

❖ Coastal Beaches

- Beaches are ever-changing and rely on sediment eroded from the land to replenish sand lost by coastal currents.
- Dams often trap the sediment and rob beaches of their natural replenishing sand.
- Protecting coastlines may require measures to ensure a steady stream of sediment from rivers and a hands-off policy toward building levees to prevent beach erosion.

CH13-30

❖ Wild and Scenic Rivers

- Rivers are a source of recreation, but they provide many other benefits.
- Unfortunately, for many years people have viewed them solely as a valuable source of water that should be dammed.
- Many rivers or segments of rivers in the U.S. have been protected from development by the Wild and Scenic Rivers Act.
- Designating a river for protection is often fraught with controversy.

CH13-31

❖ Wild and Scenic Rivers (cont.)

- Fortunately, in the U.S., the demand for new water projects has declined.
- This is largely because of:
  - a cutback in federal subsidies for such projects
  - the realization that alternative sources of water, such as water conservation, are far cheaper and much better from an environmental perspective.

CH13-32

Notes

Notes

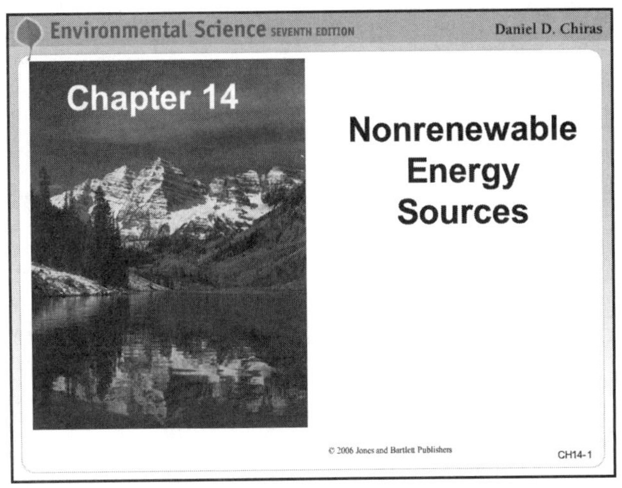

14.1 Energy Use: Our Growing Dependence on Nonrenewable Fuels

- U.S. and Canadian Energy Consumption
  - Energy use in the U.S. has shifted considerably over the years.
  - Today, the U.S. depends on a variety of fuel sources.
  - Fossil fuels provide the bulk of the energy.
  - Industry and business consume the majority of the fuel.
  - Transportation is another major energy consumer.

CH14-2

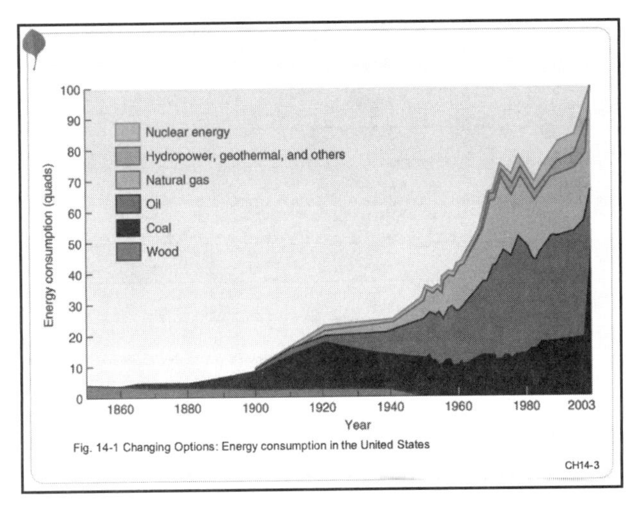

Fig. 14-1 Changing Options: Energy consumption in the United States

Notes

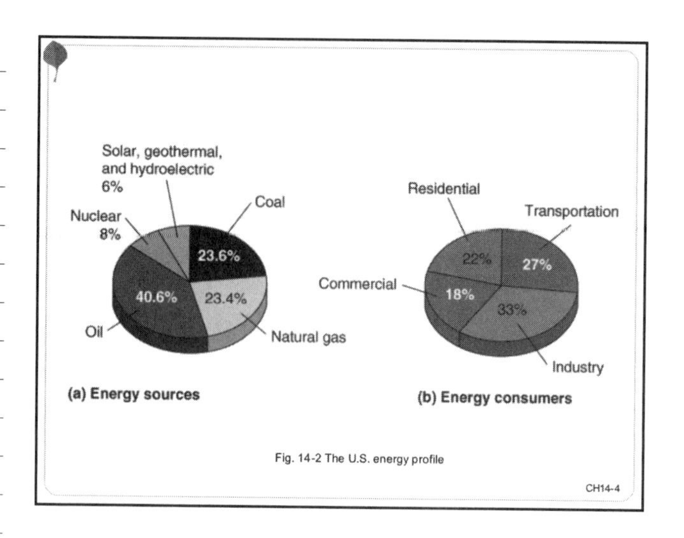

Fig. 14-2 The U.S. energy profile

CH14-4

❖ Global Energy Consumption

- Like the U.S., most MDCs rely primarily on fossil fuels.
- LDCs depend on fossil fuels as well, but they also receive a substantial amount of energy from various renewable fuels, especially biomass.
- Americans make up a small portion of the world's population but account for a very large percentage of the total energy consumption.

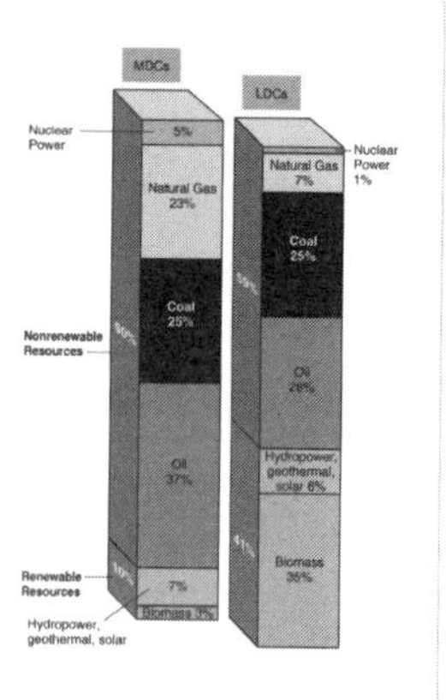

Fig. 14-3 Global energy use

CH14-5

14.2 What is Energy?

❖ Energy comes in many forms.

- Energy can be renewable or nonrenewable
- Energy can be converted from one form to another
- Energy conversion allows us to use energy
- Energy can neither be created nor destroyed
- No energy conversion is 100% efficient

CH14-6

14.3 Fossil Fuels: Analyzing Our Options

- Energy does not come cheaply.
- In addition to the economic costs, society pays a huge environmental price for its use of nonrenewable energy in damage to the health of its people and to the environment.
- These impacts arise at every phase of energy production.
- The most significant impacts arise from extraction and end use.

CH14-7

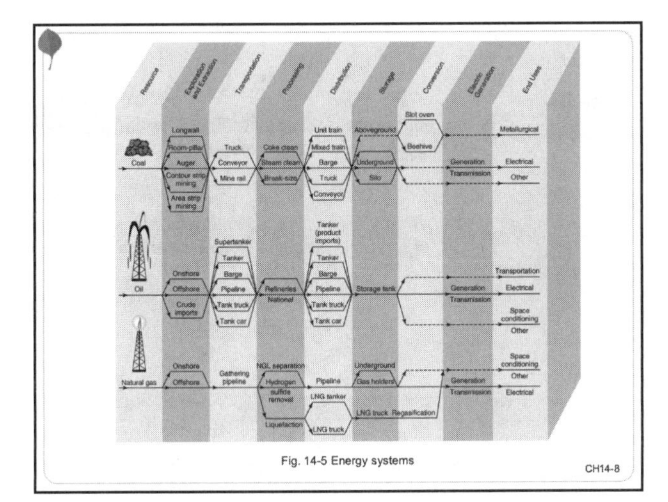

Fig. 14-5 Energy systems

CH14-8

Crude Oil

- Oil is extracted from deep wells on the seafloor and on land.
- It is often found in association with natural gas.
- After it is extracted, crude oil is heated and distilled to separate the useful fuel and nonfuel by-products.
- The major impacts of the oil energy system come from oil spills and from combustion of oil and its by-products.

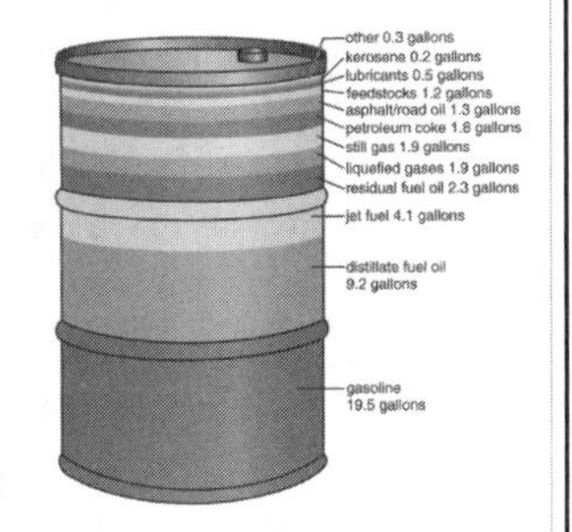

Fig. 14-6 Products from a 44-gallon barrel of oil

CH14-9

Notes

Notes

❖ Natural Gas

- Natural gas is a combustible gas extracted from deep onshore and offshore wells.
- Natural gas is used to heat homes, cook food, and heat water.
- It is also used by factories and some power plants.
- It is easy to transport within countries or from one country to its terrestrial neighbors.
- Natural gas is a clean fuel compared to oil, oil by-products, and coal.

CH14-10

❖ Coal

- Coal is the most abundant fossil fuel, but its extraction and use create enormous social, economic, and environmental costs.
- Coal is removed by surface and underground mines, both of which create many environmental impacts.
- Surface mining is especially damaging to the environment.
- Proper controls can greatly reduce impacts and mine reclamation can help companies to return lands to their original use.

CH14-11

❖ Coal (cont.)

- Underground coal mines present health and safety hazards for miners.
- Many miners have been killed or injured by accidents in the past 60 years.
- Many others suffer from black lung disease, a crippling degenerative lung disease, as a result of exposure to coal dust.

CH14-12

❖ Coal (cont.)

- One of the biggest environmental problems from coal mining results from the release of sulfuric acid from abandoned underground mines, which poisons thousands of miles of streams in the eastern U.S.

Fig. 14-10 Acid mine drainage

CH14-13

❖ Coal (cont.)

- Electric power plants are the major consumer of coal in the U.S.
- Pollution controls reduce the amount entering the atmosphere.
- The large number of power plants and the enormous quantities of coal burned still result in massive amounts of air pollution.

CH14-14

❖ Coal (cont.)

- Several air pollutants contribute to two of the world's most pressing environmental problems:
  - acid deposition
  - global climate change
- Pollutants captured by pollution control devices end up as solid waste, much of which is buried in landfills.

CH14-15

Notes

Notes

❖ Oil Shale

- Oil shale is a sedimentary rock containing an organic material (kerogen) that can be extracted from the rock by heating.
- In a liquid state, this thick, oily substance, called shale oil, can be refined to make gasoline and a host of other chemical by-products.
- Although it is abundant, oil shale is economically and environmentally costly to produce and is currently an insignificant source of fuel.

CH14-16

❖ Tar Sands

- Tar sands are sand deposits impregnated with a petroleum-like substance known as bitumen.
- Although there are large deposits, they are economically and environmentally costly to extract.

CH14-17

❖ Coal Gasification and Liquefaction

- Coal can be converted to gaseous and liquid fuels to replace oil and natural gas, but the processes are energy-intensive and produce much pollution.

CH14-18

14.3 Fossil Fuels: Meeting Future Demand

- Oil: The End Is Near
  - Global oil supplies appear to be quite large.
  - When one calculates how long they will last (based on historical increases in energy consumption), it appears that oil supplies may last only 20 to 40 years.
  - Long before the last drop of oil is removed, shortages will appear, with serious social and economic repercussions.

CH14-19

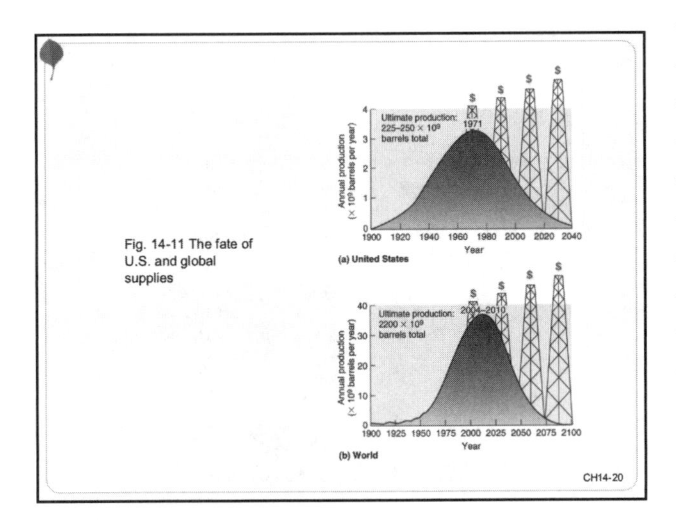

Fig. 14-11 The fate of U.S. and global supplies

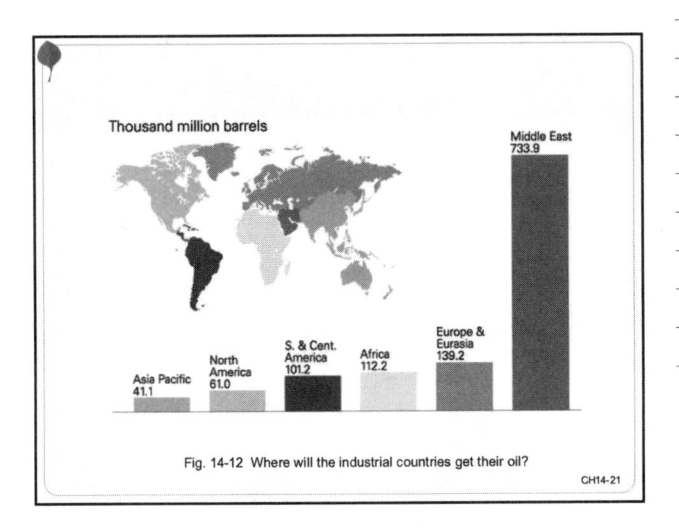

Fig. 14-12 Where will the industrial countries get their oil?

Notes

Notes

❖ Natural Gas: A Better Outlook

- Natural gas is a clean and rather abundant fossil fuel that could provide a source of energy as the world makes the transition to a more sustainable energy supply system.

US production peaked in 1973

Production generally flat since 1994

Tcf Natural gas (thousands)

Year

Fig. 14-14 U.S. marketed natural gas production 1930-2004

CH14-22

❖ Coal: The Brightest Outlook, the Dirtiest Fuel

- Coal supplies in the U.S. and many other countries are abundant, but evidence suggests that coal combustion is an environmentally unsustainable fuel.

CH14-23

14.4 Nuclear Energy

❖ Atoms: The Building Blocks of Matter

- Atoms are the building blocks of all elements.
- All atoms consist of positively charged nuclei, containing protons and neutrons, surrounded by large electron clouds in which the negatively charged electrons are found.

CH14-24

❖ Radioactive Atoms

- Elements are the purest form of matter.
- Although all atoms of a given element are the same type, some atoms contain slightly more neutrons.
- Those atoms with excess neutrons tend to be unstable and often emit radiation.

CH14-25

❖ Radiation

- Physicists have identified three major types of ionizing radiation released by radioactive atoms:
  - alpha particles
  - beta particles
  - gamma rays
- They vary in mass and ability to penetrate.
- All can cause electrons to be removed from molecules they strike.
- The X-ray is an artificially produced form of radiation that resembles the gamma ray.

CH14-26

❖ Understanding Nuclear Fission

- Uranium atoms that undergo fission release additional neutrons, causing additional fission and heat.
- The chain reaction in a nuclear reactor is kept from running rampant by:
  - bathing the reactor core with water
  - using control rods
  - maintaining the proper concentration of the fuel in the fuel rods

CH14-27

Notes

Notes

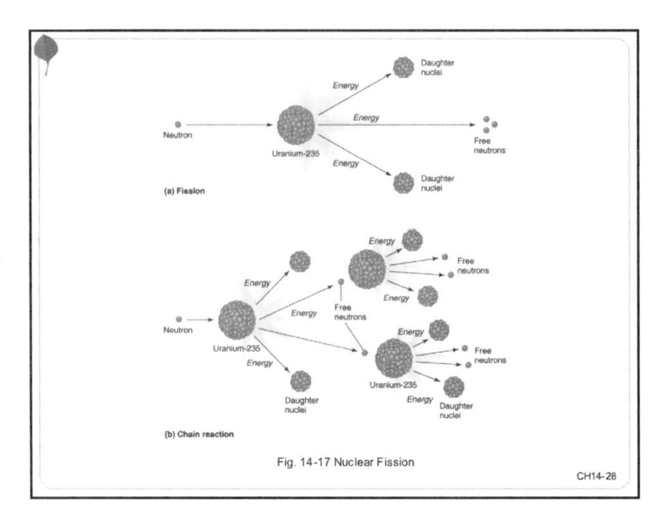

Fig. 14-17 Nuclear Fission

CH14-28

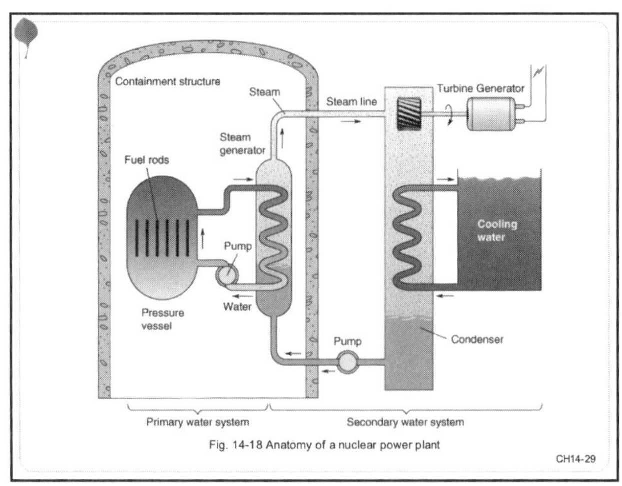

Fig. 14-18 Anatomy of a nuclear power plant

CH14-29

❖ Nuclear Power: The Benefits

- Nuclear power has many redeeming qualities.
- Although it is the most expensive of the major sources of electricity, it fits well into the established electrical grid and produces very little air pollution.

CH14-30

❖ Nuclear Power: The Drawbacks

- Interest in nuclear power has declined substantially because of major problems, among them:
  - questions over reactor safety
  - unresolved waste disposal issues
  - low social acceptability
  - high costs

CH14-31

❖ Nuclear Power: The Drawbacks (cont.)

- The effects of radiation on human health depend on many factors, such as:
  - amount of radiation
  - length of exposure
  - type of radiation
  - half-life of the radionuclide
  - age of the individual
  - the part of the body exposed

CH14-32

❖ Nuclear Power: The Drawbacks (cont.)

- High-level exposure to radiation leads to immediate death in the highest doses and radiation sickness at lower ones.
- Radiation sickness is survivable, but victims of it often suffer from serious delayed effects, including cancer and sterility.
- Pregnant women suffer from miscarriages.

CH14-33

Notes

Notes

❖ Nuclear Power: The Drawbacks (cont.)

- Scientific studies show that exposure to low-level radiation has a measurable effect on human health, the most common effect being cancer.
- Radionuclides often accumulate within tissues, where they irradiate cells.
- Thus, seemingly low levels in the environment may become dangerously concentrated in certain regions in the body.

CH14-34

❖ Nuclear Power: The Drawbacks (cont.)

- Several major accidents at nuclear power plants have raised awareness of the potential damage a small mechanical or human error might cause.
- Estimates suggest that many additional accidents are bound to occur in the future, with costly social, economic, and environmental impacts.

CH14-35

❖ Nuclear Power: The Drawbacks (cont.)

- Nuclear power has a long history in the U.S.
- In the early years, many wastes were carelessly disposed of, whereas others remain stockpiled at nuclear power plants.
- The U.S. nuclear industry still lacks acceptable means of disposing of the high-level wastes produced by reactors and uranium mills.

CH14-36

❖ Nuclear Power: The Drawbacks (cont.)

- Nuclear power has become a socially unacceptable form of electricity in part because of high costs during all phases of operation, from construction to operation, repair, and retirement.
- Countries with nuclear power plants can develop atomic bombs from waste products.

CH14-37

❖ Breeder Reactors

- Breeder reactors are designed to produce energy and radioactive fuel from abundant nonfissile (nonfissionable) uranium-238
- They could greatly increase the supply of nuclear fuels.
- Unfortunately, breeder reactors are costly to build and take 30 years to produce as much radioactive fuel as they consume.

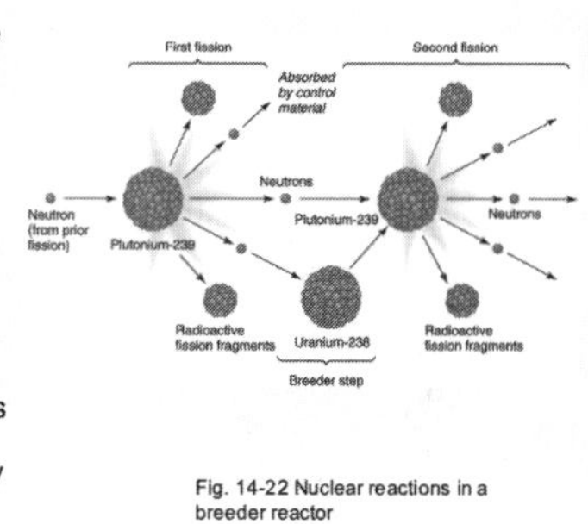

Fig. 14-22 Nuclear reactions in a breeder reactor

CH14-38

❖ Nuclear Fusion

- Enormous amounts of energy are produced when small atoms fuse.
- Fusion reactors could be powered by plentiful fuel sources and could supply energy for many millions of years.
- However, technical problems, costs, and environmental hazards present major obstacles to the commercial development of this form of energy.

CH14-39

Notes

Notes

14.5 General Guidelines for Creating a Sustainable Energy System

- There are many energy options, but not all are sustainable.
- Creating a sustainable energy future will require a careful analysis of options for such factors as:
  - net energy yield
  - specific needs
  - efficiency
  - environmental impacts
  - abundance and renewability
  - affordability

CH14-40

14.6 Establishing Priorities

❖ Short-term goals (within 10-20 years)

- In the near term, efforts are needed to:
  - improve the efficiency of all energy-consuming technologies
  - find sustainable alternatives to coal, crude oil, and their derivatives

❖ Long-term goal (within 50 years)

- In the long term we must find sustainable replacements for natural gas.

CH14-41

Chapter 15: Foundations of a Sustainable Energy System: Conservation and Renewable Energy

Notes

Environmental Science SEVENTH EDITION Daniel D. Chiras

Chapter 15

Foundations of a Sustainable Energy System

Conservation and Renewable Energy

CH15-1

15.1 Energy Conservation: Foundation of a Sustainable Energy System

- Energy is used wastefully in virtually all nations.
- Excessive waste is a sign of an unsustainable technology.

Table 15-1
Oil Consumed per Unit of National Output, Selected Countries, 2000

Country	Tons of Oil Equivalent per $1000 (US)
Canada	0.40
United States	0.34
Germany	0.26
France	0.25
United Kingdom	0.24
Japan	0.19
Switzerland	0.17
Italy	0.16

CH15-2

❖ Economic and Environmental Benefits of Energy Conservation

- Energy-efficiency measures can reduce the world's dependence on fossil fuels.
- Energy-efficiency measures are easy to implement and cost a fraction of what new energy supplies do, while providing many environmental benefits.

CH15-3

Notes

❖ Energy-Efficiency Options

- There's no shortage of easy, economical ways to save energy.
- These include simple behavioral changes, such as turning off lights when one leaves a room, as well as many energy-efficient technologies, such as:
  - compact fluorescent light bulbs
  - cogeneration facilities
- Energy savings result in substantial social, economic, and environmental benefits.

CH15-4

❖ The Potential of Energy Efficiency

- Because energy is used so inefficiently, huge cuts in energy demand can be made by applying efficiency measures.
- This shouldn't affect the level of services we receive.
- Much of our future energy demand can be met by freeing up energy currently wasted in three areas:
  - transportation
  - buildings
  - industry/business

CH15-5

❖ The Potential of Energy Efficiency (cont.)

- Extremely energy-efficient vehicles are currently available.
- Many improvements in vehicles could increase efficiency even more, greatly cutting transportation energy consumption.

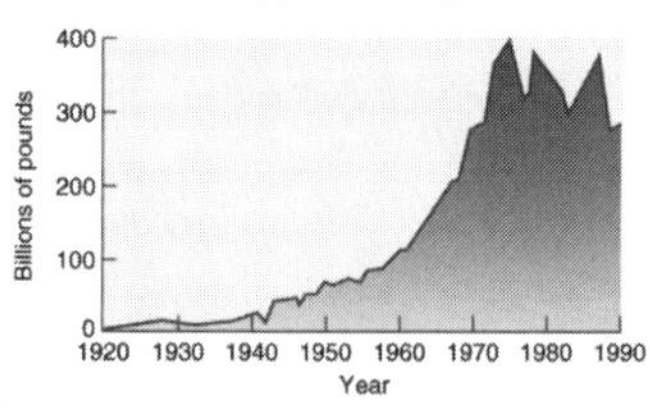

Fig 15-3 Energy consumption in the transportation sector

CH15-6

❖ The Potential of Energy Efficiency (cont.)

- Energy-efficiency measures in buildings can result in substantial energy savings in:
  - Heating
  - Cooling
  - Lighting
  - Appliances and electronic equipment
- Short-term thinking and economics often prevent investments in measures to cut energy demand in these areas.

CH15-7

❖ The Potential of Energy Efficiency (cont.)

- Industrial energy consumption accounts for a major portion of the world's energy demand.
- It could be cut substantially, making companies more profitable and competitive.

Fig 15-4 Energy vs. GNP

CH15-8

❖ Promoting Energy Efficiency

- There are many ways to use energy much more efficiently.
- There are also many ways to promote this strategy, including:
  1. education
  2. taxes on fossil fuels
  3. "feebate" systems (a tax on those who choose energy-inefficient options and rebates to those who opt for energy-efficient technologies)
  4. government-mandated efficiency programs
  5. changes in pricing
  6. least-cost planning

CH15-9

Notes

Notes

❖Roadblocks to Energy Conservation

- Many roadblocks stand in the way of energy efficiency, including:
  - the illusion of abundance
  - federal subsidies that underwrite fossil fuels' true costs
  - higher initial costs for some energy-efficient products
  - powerful political forces
- Despite this, energy efficiency is becoming a popular strategy.

CH15-10

15.2 Renewable Energy Sources

- Renewable energy will very likely become a major source of energy in the future.
- The transition to a renewable energy future has already begun in some nations.

CH15-11

❖ Solar Energy Options

- Solar energy is considered a renewable energy source, but it is really finite.
- Nonetheless, because it is so abundant and clean, it will very likely be a major contributor to future world energy supplies.
- Buildings can be designed to capture solar energy to provide space heat.
- Properly designed structures can derive 100% of their heat from the sun.

CH15-12

Notes

❖ Solar Energy Options (cont.)

- Active solar systems generally employ rooftop panels that collect heat from sunlight and store it in water or some other medium.
- This solar energy can then be used to heat domestic hot water or to heat the interior of the building.

Fig 15-8 Active solar heating system

CH15-13

❖ Solar Energy Options (cont.)

- Photovoltaics are thin wafers of material such as silicon that emits electrons when struck by sunlight, creating electricity.
- Although photovoltaics are costly, prices are falling.
- Solar thermal electric facilities heat water using sunlight.
- Steam from this fairly cost-competitive process is used to generate electricity.

Fig 15-10a Photovoltaic panels

CH15-14

❖ Solar Energy Options (cont.)

- Solar energy technologies are well developed.
- Their advantage over other forms of energy production is that they rely on a free, abundant fuel and are relatively clean systems to operate.
- Although some systems are economically competitive, others are still fairly costly.
- Storing energy from intermittent sunlight remains one of their major drawbacks.

CH15-15

Notes

❖ Wind Energy

- Winds are produced by solar energy and can be used to generate electricity or to perform work directly, such as pumping water.
- Wind energy is clean, abundant, and fairly inexpensive, especially when one includes its low environmental costs.
- Wind energy could provide a significant percentage of our future energy demand.
- However, because winds are often intermittent, backup systems and storage are necessary.

CH15-16

Table 15-2

Land Use of Selected Electricity-Generating Technologies, United States

Technology	Land Occupied (square meters per gigawatt-hour, for 30 years)
Coal[1]	3642
Solar thermal	3561
Photovoltaics	3237
Wind[2]	1335
Geothermal	404

Source: From L.R. Brown, C. Flavin, and S. Postel (1991). *Saving the Planet: How to Shape an Environmentally Sustainable Global Economy.* New York: Norton.

[1]Includes coal mining.

[2]Land actually occupied by turbines and service roads.

CH15-17

Table 15-3

Present and Estimated Electrical Generation Costs[1]

Source	Cents per Kilowatt-Hour
Nuclear	8–12
Coal	5–7
Gas and oil	6–9
Hydroelectric	3–6
Wind	5–8
Geothermal	4.5–5.5
Photovoltaic	22
Solar thermal	8–12
Biomass	5

Sources: Public Citizen Critical Mass Energy Product and Worldwatch Institute.

[1]Generation costs are costs to companies.

CH15-18

Notes

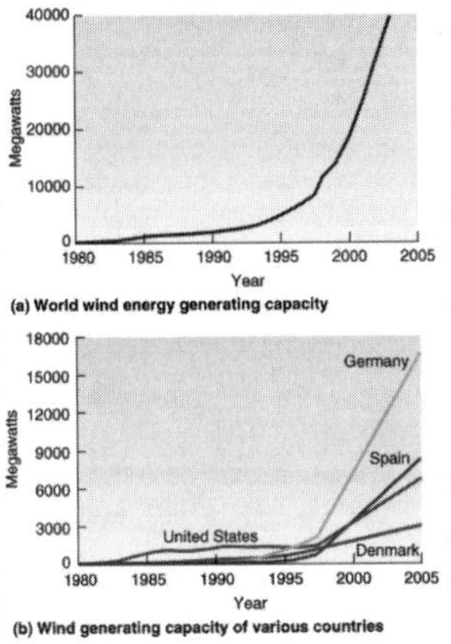

40000
30000
20000
10000
0
Megawatts
1980 1985 1990 1995 2000 2005
Year
(a) World wind energy generating capacity

18000
15000
12000
9000
6000
3000
0
Megawatts
Germany
Spain
United States
Denmark
1980 1985 1990 1995 2000 2005
Year
(b) Wind generating capacity of various countries

Fig. 15-14 Wind energy

CH15-19

❖ Biomass

- Biomass is organic matter such as wood or crop wastes that can be burned or converted into gaseous or liquid fuels.
- It is a common fuel source in most developing nations but supplies only a fraction of the needs of people in the developed nations.

CH15-20

❖ Biodiesel

- Biodiesel is a renewable fuel made from an assortment of vegetable oils and a methanol-lye mixture
- There are many potential sources for biodiesel

CH15-21

Notes

❖ Hydroelectric Power

- Hydroelectric power is renewable and operates relatively cleanly, but dams and reservoirs have an enormous impact on the environment.

- Although potential hydropower sources are enormous, they can be far from settlements, and developing them would cause serious environmental damage.

CH15-22

❖ Geothermal Energy

- Geothermal energy is a renewable resource created primarily from magma, molten rock beneath the crust.
- Geothermal energy is used to generate electricity and to heat structures.
- It is a major source of energy in some countries.

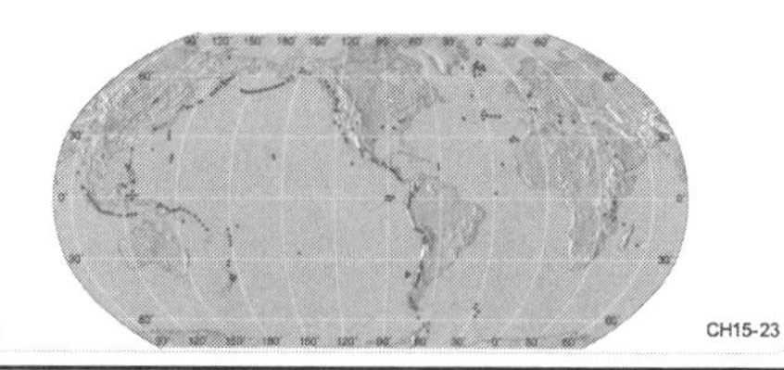

Fig. 15-16 Global geothermal resources

CH15-23

❖ Hydrogen Fuel

- Hydrogen may become an important fuel in the future.
- Hydrogen can be produced by passing electricity through water, a renewable resource.
- When hydrogen burns, it produces water vapor.
- Fuel cells use hydrogen, either from water or organic fuels, to produce electricity.
- The electricity can be used to power cars, and several manufacturers are actively pursuing this option.

CH15-24

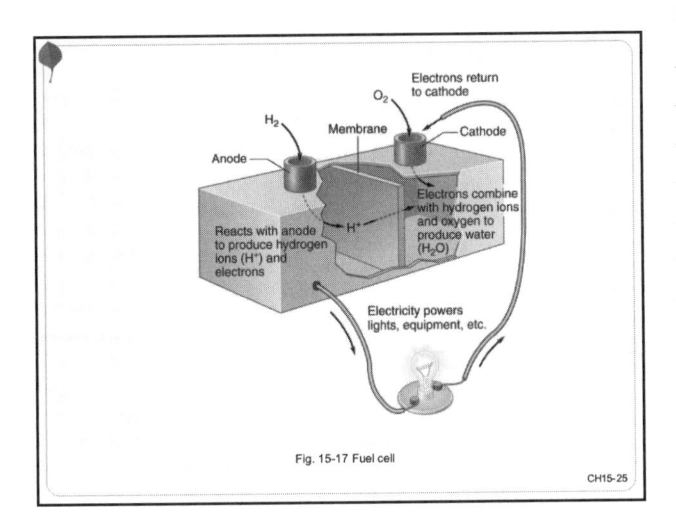

Fig. 15-17 Fuel cell

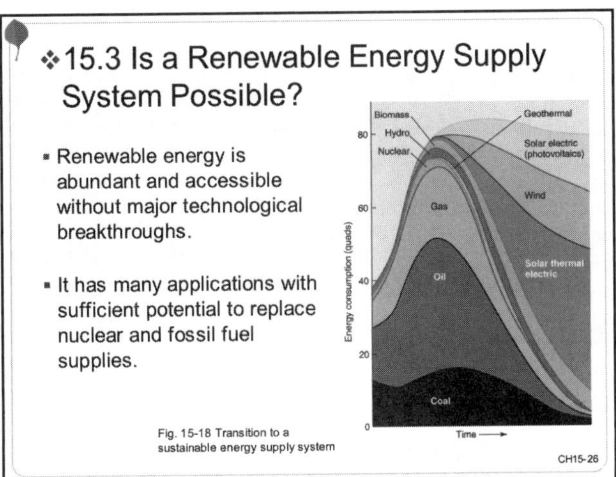

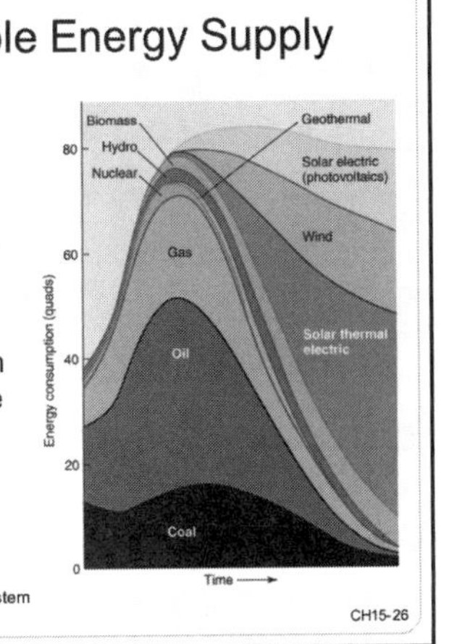

Fig. 15-18 Transition to a sustainable energy supply system

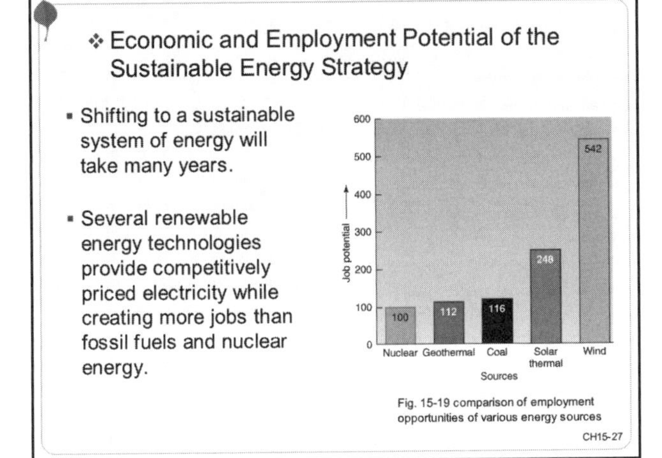

Fig. 15-19 comparison of employment opportunities of various energy sources

Notes

Notes

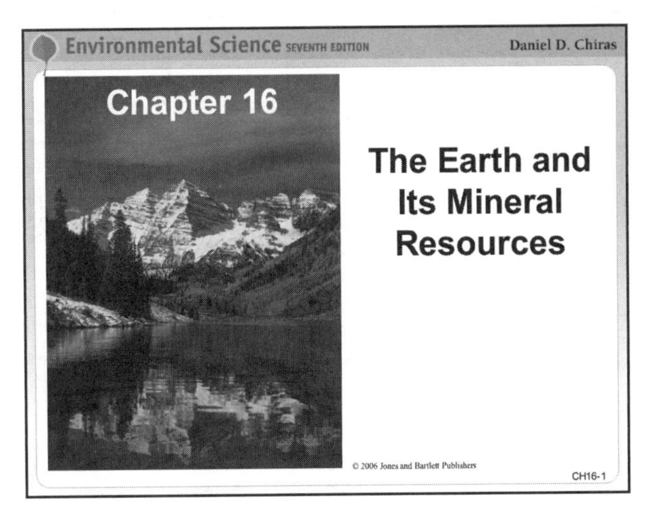

16.1 The Earth's Mineral Riches

- The Earth contains many valuable nonfuel minerals.
- Those metal-yielding minerals found in concentrated deposits called ores are mined and processed to produce metals.

CH16-2

❖ Mineral Resources and Society

- More than a hundred nonfuel minerals are traded in the world market.
- These materials, worth billions of dollars to the world economy, are vital to industry, agriculture, and our own lives.

Table 16-1
Estimated World Production of Selected Minerals, 2003

Mineral	Production* (thousand tons)
Metals	
Aluminum	27,700
Copper	13,600
Manganese	8,200
Zinc	9,010
Chromium	15,500
Lead	2,950
Nickel	1,400,000
Tin	207,000
Molybdenum	125,000
Titanium	5,300
Silver	18,800
Mercury	1,530
Platinum-group metals	205,000
Gold	2,590
Nonmetals	
Sand and gravel	110,000,000
Clays	36,000
Salt	210,000
Phosphate rock	137,000
Lime	120,000
Gypsum	104,000
Soda ash	38,000
Potash	28,000

*All data exclude recycling.

CH16-3

Notes

❖ Who Consumes the World's Minerals?

- The more developed nations consume the bulk of the world's minerals.
- As the less developed nations' economies expand, they will use a larger share of the world's mineral supplies.

CH16-4

❖ Import Reliance

- Many more developed nations such as the U.S. import a large percentage of their minerals.
- Some minerals come from politically volatile areas.
- Disruptions in the supplies from these nations could result in considerable economic hardship.
- More developed nations typically stockpile minerals to protect themselves against such possibilities.

CH16-5

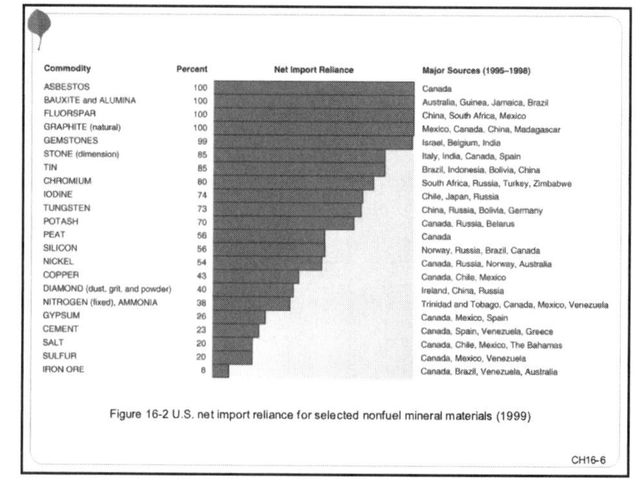

Figure 16-2 U.S. net import reliance for selected nonfuel mineral materials (1999)

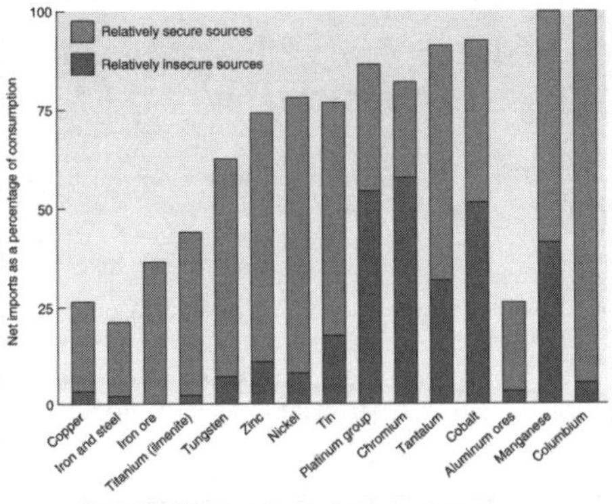

Relatively secure sources
Relatively insecure sources

Net imports as a percentage of consumption

100, 75, 50, 25, 0

Copper, Iron and steel, Iron ore, Titanium (ilmenite), Tungsten, Zinc, Nickel, Tin, Platinum group, Chromium, Tantalum, Cobalt, Aluminum ores, Manganese, Columbium

Figure 16-3 U.S. mineral imports from politically unstable countries

CH16-7

Notes

❖ Will There Be Enough?

- Most minerals are present in adequate quantities or can be replaced by various substitutes.
- Some—about 18 minerals we use—are in danger of falling into short supply within the near future.

CH16-8

16.2 Environmental Impacts of Mineral Exploitation: A Brief Overview

- The mineral production/consumption cycle produces extraordinary environmental impacts.
- The most noticeable occur in the mining and smelting phases.

CH16-9

Notes

16.3 Supplying Mineral Needs Sustainably

- Creating a Sustainable System of Mineral Production
  - Implementing the operating principles of sustainability—conservation, recycling, and restoration—can help human society create a more sustainable system of mineral production.

CH16-10

- Creating a Sustainable System of Mineral Production (cont.)
  - Recycling metals provides materials needed to manufacture goods at a fraction of the environmental cost of producing them from raw materials.
  - This is in large part because using already processed and refined materials saves energy.
  - Recycling can help to extend mineral supplies.

CH16-11

- Conservation—Decreasing Product Size, Increasing Product Durability
  - Using less material by downsizing products or by making them more durable will help stretch limited supplies.
  - These steps will help promote efficiency and conservation:
    - reforming unsustainable laws
    - removing subsidies on raw materials
    - giving financial incentives to companies that use resources efficiently and incorporate recycled materials in their products

CH16-12

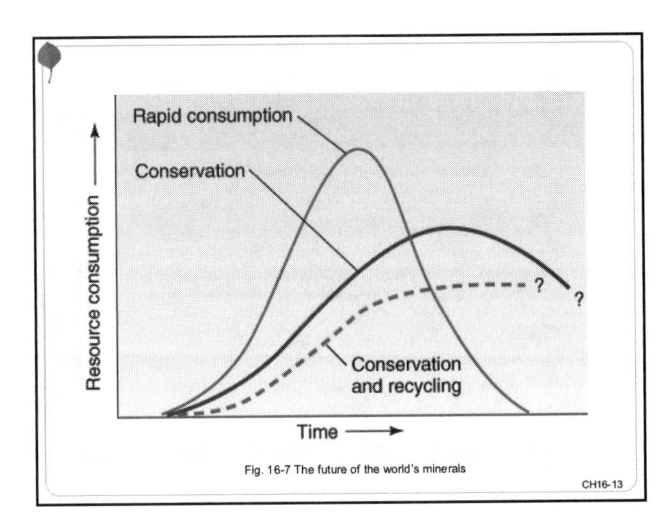

Fig. 16-7 The future of the world's minerals

CH16-13

❖ Restoration and Environmental Protection

- Meeting mineral demands sustainably will require efforts to minimize environmental damage from mining and other operations.
- Tougher laws may be needed.
- Laws that exempt the mining industry from environmental protection may need to be changed.
- Better efforts to restore the damage to natural systems caused by mineral production are needed in many countries.

CH16-14

16.4 Expanding Reserves

- Future demand cannot all be satisfied by recycling and conservation efforts.
- Some new minerals must be mined.

CH16-15

Notes

Notes

❖ Rising Prices, Rising Supplies

- Many factors determine the size of mineral reserves.
- One of the most important is the price.
- Reserves tend to expand as prices rise, because companies are willing to spend more to develop lower-grade ores.
- Ultimately all mineral resources are finite.

CH16-16

❖ Technological Advances Expand Reserves

- Technological improvements make it feasible to mine less concentrated ores, which helps expand reserves.

❖ Factors That Reduce Supplies

- high labor costs
- interest rates
- energy costs
- environmental protection costs

CH16-17

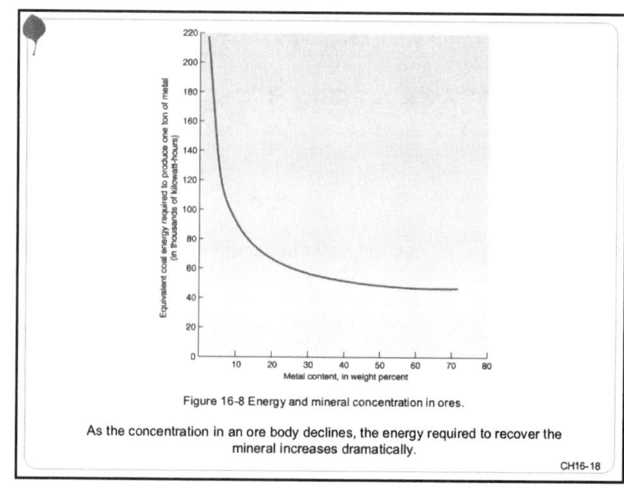

Figure 16-8 Energy and mineral concentration in ores.

As the concentration in an ore body declines, the energy required to recover the mineral increases dramatically.

CH16-18

❖ Minerals from the Sea

- Mineral-rich nodules are found on the ocean's floor.
- Although they appear to be economically feasible to mine, little is known about the environmental impact.
- Questions of ownership also plague their exploitation.

CH16-19

16.5 Finding Substitutes

❖ Substitution of one resource for another that has become economically depleted has been a useful strategy in the past, but it may not work in all cases.

❖ Substitutes have limits, and some materials have no suitable alternatives.

CH16-20

16.6 Personal Actions

❖ Individual action is vital to building a sustainable future.

❖ Steps you can take:

- buying durable products
- recycling
- choosing recycled materials

CH16-21

Notes

Notes

Environmental Science SEVENTH EDITION Daniel D. Chiras

Chapter 17

Creating Sustainable Cities, Suburbs, and Towns

Principles and Practices of Sustainable Community Development

CH17-1

17.1 Cities and Towns as Networks of Systems

- Cities and towns consist of numerous systems, such as energy, housing, and transportation.
- Many experts think these systems are largely unsustainable.
- Making our living environment sustainable will require us to redesign human systems to better fit within the natural systems that support us.

CH17-2

❖ The Invisibility of Human Systems

- Most efforts to solve environmental problems have focused on treating symptoms rather than on rethinking and revamping the systems at the root of the problems.
- Most people are unaware of the systems that support our lives until they break down.

CH17-3

Notes

❖ Performance versus Sustainability: Understanding a Crucial Difference

- Just because a system such as energy or manufacturing appears to be functioning well does not mean it is sustainable in the long run.

CH17-4

❖ Why Are Human Systems Unsustainable?

- Human systems are unsustainable because they exceed the carrying capacity of the Earth. They:
  - produce pollution in excess of the planet's ability to absorb it
  - use renewable resources faster than they can be replenished
  - deplete nonrenewable resources

CH17-5

❖ The Challenge of Creating Sustainable Cities and Towns

- Two challenges face existing communities:
  - revamping existing infrastructure
  - building new infrastructure in as sustainable a manner as possible

Fig. 17-3 Solar home

CH17-6

Notes

17.2 Land-Use Planning and Sustainability

- Land-use planning helps cities establish the locations of various structures and activities.
- They help keep incompatible uses apart.
- As conceived and practiced in most places, it doesn't do much for sustainability.

CH17-7

❖ Sustainable Land-Use Planning: Ending Sprawl

- Sustainable land-use planning and development seek to optimize land use and minimize the loss of economically and ecologically important lands.
- They offer other benefits as well, including:
  - more efficient mass transit
  - reduced air pollution
  - reductions in the cost of providing water, sewage, and other services

CH17-8

❖ Sustainable Land-Use Planning: Ending Sprawl (cont.)

- Dispersed development, or **urban sprawl**, is the most environmentally and economically unsustainable form of urban/suburban development.
- Compact development is a denser settlement pattern that offers many environmental benefits, such as reduced land use and air pollution and increased efficiency of mass transit.

CH17-9

Notes

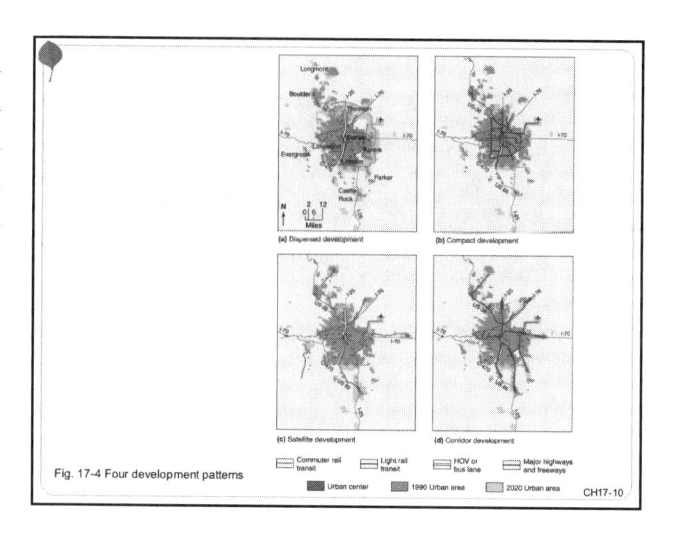

Fig. 17-4 Four development patterns

❖ Sustainable Land-Use Planning: Ending Sprawl (cont.)

- Corridor and satellite development, concentrating development along major traffic routes or in distant communities connected to the metropolitan area, are more desirable than urban sprawl but less sustainable than compact development.

CH17-11

❖ Land-Use Planning and Building

- Many steps can be taken when building homes to protect valuable ecological assets.
- These can save developers money and reduce cost.
- Statewide and Nationwide Sustainable Land-Use Planning
- Many states and nations have land-use planning that minimizes urban sprawl.

CH17-12

❖ Beyond Zoning

- Many methods can help promote sustainable land-use patterns, including:
  - zoning
  - differential tax rates
  - purchase of development rights
  - making growth pay its own way
  - open space acquisition

CH17-13

❖ Land-Use Planning in the Less Developed Nations

- Land-use planning and land reform are also essential to creating sustainable land-use patterns in the developing nations.

CH17-14

17.3 Shifting to a Sustainable Transportation System

❖ Automobiles are a major component of the global transportation system.

❖ Problems associated with their use that are likely to help stimulate a shift to a more sustainable transportation system include:

- declining oil supplies
- congestion in urban areas
- regional air pollution problems
- global climate change

CH17-15

Notes

Notes

❖ Phase 1: The Move toward Efficient Vehicles and Alternative Fuels

- More efficient cars are part of the first phase of the transition to a sustainable energy system.
- Improvements in engines and automobile aerodynamics are key elements of this effort.
- Other keys to the success of these efforts are computer systems that :
  - operate cars automatically
  - monitor traffic and signal congestion
  - permit designers to create more efficient highways

CH17-16

❖ Phase 1: The Move toward Efficient Vehicles and Alternative Fuels (cont.)

- Aircraft manufacturers have made much more impressive strides in improving fuel efficiency.
- Alternative fuels that burn cleanly and are renewable could also help reduce many problems created by the gasoline-powered automobile.

CH17-17

❖ Phase 2: From Road and Airports to Rails, Buses, and Bicycles

- Mass transit is much more efficient than automobiles and produces much less pollution per passenger mile traveled.
- Congestion, fuel concerns, and interest in cutting pollution will all stimulate the shift to efficient mass transit in urban areas.
- More compact development patterns will help complement the move to mass transit.

CH17-18

❖ Phase 2: From Road and Airports to Rails, Buses, and Bicycles (cont.)

- In many cities, bicycles already carry a significant number of commuters.
- The bicycle could help supplement the mass transit systems of cities in the future.

Table 17-2
Relative Efficiencies of Various Modes of Transportation

Mode of Transportation	Kilojoules per Passenger Kilometer (passenger mile)[1]
Van pool	400 (640)
Rail	400 (640)
Bus	450 (720)
Car pool	650 (1040)
Automobile	1800 (2800)
Airline	3800 (6080)

Source: Worldwatch Institute.
[1]A kilojoule is 1000 joules, a unit of energy.

CH17-19

❖ Economic Changes Accompanying a Shift to Mass Transit

- A shift away from the automobile will have serious repercussions on the global economy.
- Much of the slack could be taken up by a shift to the manufacture and maintenance of alternative transportation modes such as buses.

CH17-20

Notes

Notes

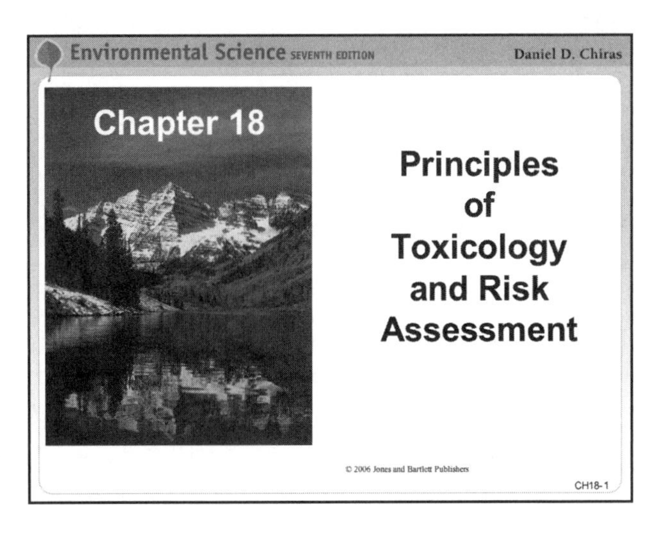

18.1 Principles of Toxicology

- Many thousands of chemicals are produced each year in industrialized nations.
- Only a small percentage pose a risk to humans.
- Vast gaps exist in our knowledge of the effects of toxic substances.
- This is largely because of the sheer number of chemicals that need to be tested and the expense of thorough testing.

CH18-2

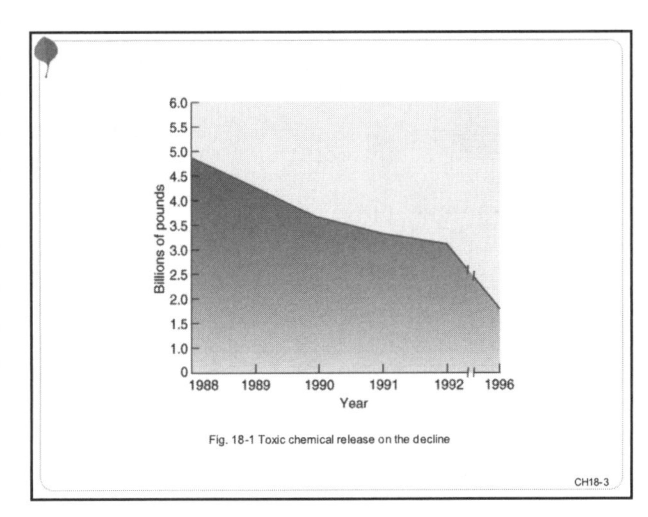

Fig. 18-1 Toxic chemical release on the decline

CH18-3

Notes

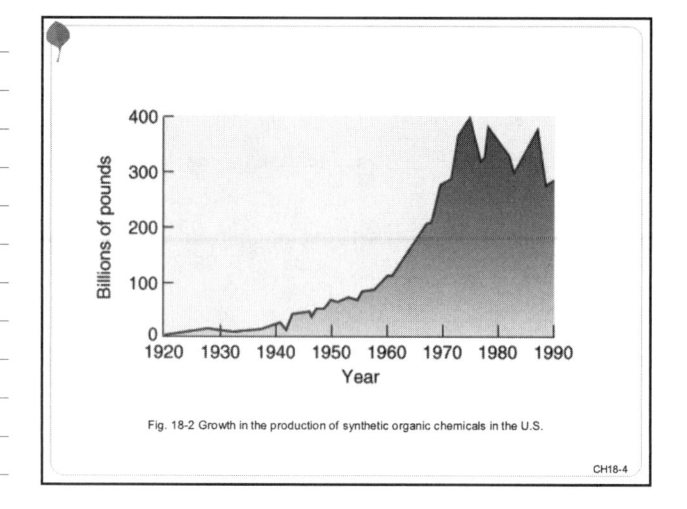

Fig. 18-2 Growth in the production of synthetic organic chemicals in the U.S.

CH18-4

❖ The Biological Effects of Toxicants

- Humans are exposed to potentially toxic substances in virtually every aspect of our lives.
- Exposure does not necessarily mean that we will be adversely affected.
- Toxic substances may produce immediate effects, ranging from slight to severe, or delayed effects, with a similar range of severity.
- Toxic substances can exert their effects locally, often at the site of exposure, or systemically (in an organ system or even throughout much of the body).

CH18-5

❖ How Toxicants Work

- Toxic substances may bind to enzymes and other important molecules, including the genetic material, DNA.
- Binding to these chemicals alters cellular function, sometimes in profound ways.

CH18-6

Notes

❖ Factors That Affect the Toxicity of Chemicals

- The toxicity of a chemical (how toxic it is) is determined by a variety of factors:
  - the route of entry
  - dose
  - length of exposure
  - age of the individual
  - sex
  - genetic makeup

CH18-7

❖ Factors That Affect the Toxicity of Chemicals (cont.)

- The effect a chemical has on the body is determined by:
  - its reactivity
  - the amount one is exposed to (known as the dose)
  - and the duration of exposure (how long one is exposed to a particular substance)

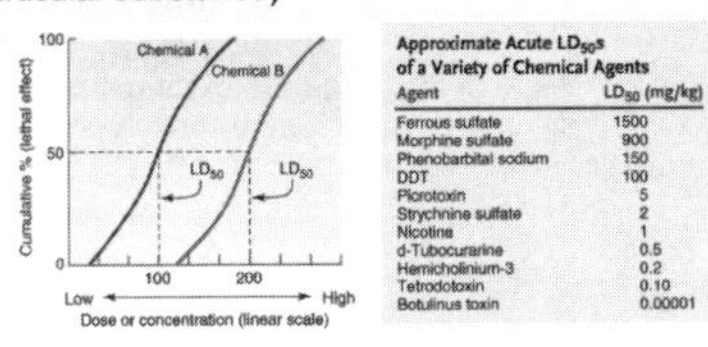

Approximate Acute LD_{50}s of a Variety of Chemical Agents

Agent	LD_{50} (mg/kg)
Ferrous sulfate	1500
Morphine sulfate	900
Phenobarbital sodium	150
DDT	100
Picrotoxin	5
Strychnine sulfate	2
Nicotine	1
d-Tubocurarine	0.5
Hemicholinium-3	0.2
Tetrodotoxin	0.10
Botulinus toxin	0.00001

Fig. 18-3 Dose-response graph for two chemicals with differing toxicities

CH18-8

❖ Factors That Affect the Toxicity of Chemicals (cont.)

- Chemicals enter the body via three major routes:
  - Inhalation
  - Ingestion
  - Dermal absorption
- Inhalation is the fastest route and dermal (skin) absorption the slowest.

CH18-9

Notes

❖ Factors That Affect the Toxicity of Chemicals (cont.)

- The age and health status of an individual are also instrumental in determining the effect of a chemical.

CH18-10

❖ Factors That Affect the Toxicity of Chemicals (cont.)

- We are typically exposed to many different toxic chemicals at different doses and for different lengths of time.
- Toxic substances may accumulate in certain tissues and organs, causing local effects.
- Some may also increase in concentration in the food chain, being the highest in top-level consumers.
- Because of these phenomena, ambient concentrations of a toxicant may be an insufficient means of predicting its toxic effects.

CH18-11

Fig. 18-7 Biological magnification

CH18-12

18.2 Mutations, Cancer, and Birth Defects

❖ Mutations

- Chemical and physical agents can alter the hereditary material in a number of ways.
- Such changes, called mutations, can occur in body cells or in reproductive cells.
- Those occurring in body cells may kill the cells or lead to uncontrolled growth, a cancer.
- Nonlethal changes in reproductive cells can be passed on to one's offspring.

CH18-13

❖ Cancer

- Cancer is the uncontrolled proliferation of cells of the body.
- Tumor cells typically develop in rapidly dividing cells such as those of the skin and often spread to other parts of the body.
- Hundreds of different types of cancer exist and many behave differently, a fact that has complicated efforts to find a cure for the disease.

CH18-14

❖ Cancer (cont.)

- Cancers can be caused by chemical, physical, and biological agents.
- Chemicals that cause cancer, called carcinogens, generally require repeated exposures over many years.
- Most carcinogens react with the growth control genes of cells, causing mutations that lead to cancer.

CH18-15

Notes

Notes

❖ Cancer (cont.)

- Evidence suggests that some other chemical carcinogens cause cancer through mechanisms that don't directly involve the DNA.
- Most chemical carcinogens are not directly mutagenic.
- They must be chemically altered by enzymes in the body to be able to react with DNA.
- This process is called biotransformation.

CH18-16

❖ Birth Defects

- Birth defects include structural and functional defects in newborns, which may be caused by chemical, physical, and biological agents called **teratogens**.

Table 18-2

Some Known and Suspected Teratogens in Humans

Known Agents	Possible or Suspected Agents
Progesterone	Aspirin
Thalidomide	Certain antibiotics
Rubella (German measles)	Insulin
	Antitubercular drugs
Alcohol	Antihistamines
Radiation	Barbiturates
	Iron
	Tobacco
	Antacids
	Excess vitamins A and D
	Certain antitumor drugs
	Certain insecticides
	Certain fungicides
	Certain herbicides
	Dioxin
	Cortisone
	Lead

CH18-17

❖ Birth Defects (cont.)

- Birth defects may occur when an embryo is exposed to a teratogen during the critical period of organ development, organogenesis.
- Chemicals vary in their effects.
- Some affect specific organ systems.
- Others affect a number of systems.

CH18-18

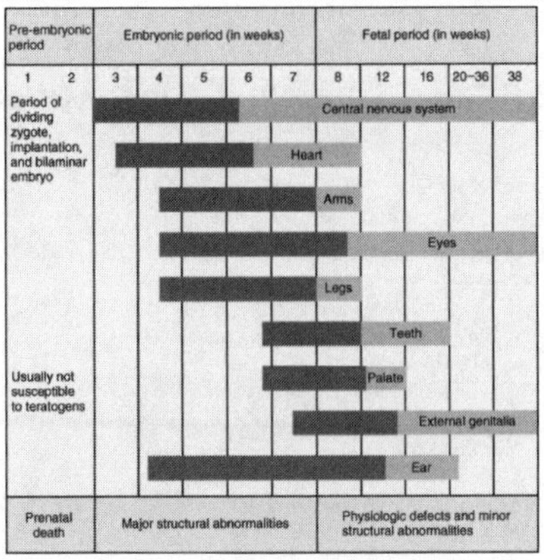

Notes

Fig. 18-10 Embryonic development and teratogenesis

CH18-19

18.3 Reproductive Toxicity

- Reproduction is a complex process, involving many steps.
- Chemical and physical agents may interrupt any of these complex processes, interfering with reproduction.

CH18-20

18.4 Environmental Hormones

- Certain pollutants in the environment enter the bodies of organisms and can alter the normal release of hormones, upsetting reproduction and other vital functions.

CH18-21

Notes

18.5 Case Studies: A Closer Look

- Asbestos: How Great a Danger?
  - Asbestos is a naturally occurring silicate mineral fiber with many practical uses.
  - Unfortunately, it produces three disorders:
    - pulmonary fibrosis (buildup of scar tissue)
    - lung cancer
    - Mesothelioma
  - It is especially dangerous to asbestos workers and to individuals who also smoke.

CH18-22

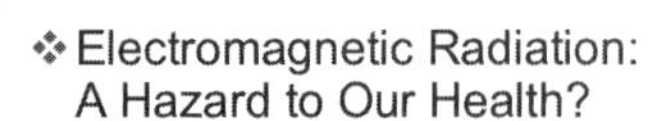

- Electromagnetic Radiation: A Hazard to Our Health?
  - Extremely low-frequency (ELF) magnetic fields are produced when electricity flows through wires.
  - Some studies suggest that these fields may increase the incidence of cancer in children living nearby.
  - To date, however, results of studies on the effects of ELF magnetic fields have yielded inconclusive results.

CH18-23

18.6 Controlling Toxic Substances: Toward a Sustainable Solution

- Many laws have been passed in the U.S. and other countries to reduce the release of toxic chemicals into the environment.
- Most of these relied on end-of-pipe controls.
- More recent efforts are aimed at preventing pollution: eliminating hazardous waste production in the first place.

CH18-24

❖ Toxic Substances Control Act

- The Toxic Substances Control Act seeks to:
  - prevent the introduction of chemicals that will be harmful to people and the environment
  - eliminate those already in use that pose an unacceptable risk

CH18-25

❖ Market Incentives to Control Toxic Chemicals

- Many efforts are under way to harness market forces, rather than impose regulations on companies, to encourage the use of nontoxic or less toxic products.
- These mechanisms promote:
  - innovation
  - freedom of choice
  - cost-saving solutions that many businesses view as an acceptable way to reduce pollution

CH18-26

❖ The Multimedia Approach to Pollution Control: An Integrated Approach

- Pollutants from factories exit via one of several avenues, such as wastewater or air pollution.
- For years, each medium has been regulated separately.
- Efforts are now under way to regulate and monitor several media simultaneously to keep companies from dumping potentially toxic substances in the least-regulated medium.

CH18-27

Notes

Notes

18.7 Risk and Risk Assessment

- Understanding and quantifying the risks posed in modern society is essential to creating socially, economically, and environmentally acceptable policies.

CH18-28

❖ Risks and Hazards: Overlapping Boundaries

- Risk arises from natural events as well as human activities.
- Human activities profoundly influence natural hazards.

CH18-29

❖ Three Steps in Risk Assessment

- Risk assessment is a technique that helps us
  1. identify risks
  2. determine the probability or likelihood of their occurrence
  3. assess the potential severity of the effects
     - the potential economic, health, social, and environmental costs

CH18-30

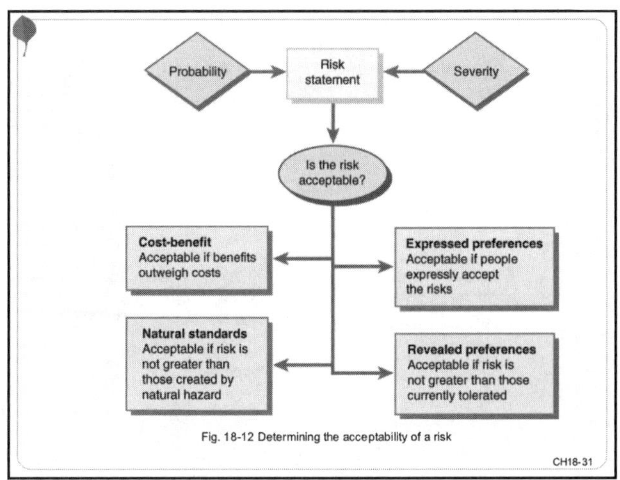

Fig. 18-12 Determining the acceptability of a risk

CH18-31

❖ Three Steps in Risk Assessment (cont.)

- Assessing the risk of a toxic substance is difficult because
  - so little is known about the thousands of chemicals
  - much of the work is done on laboratory animals, so may not apply to humans

Fig. 18-13 Hypothetical dose-response curves

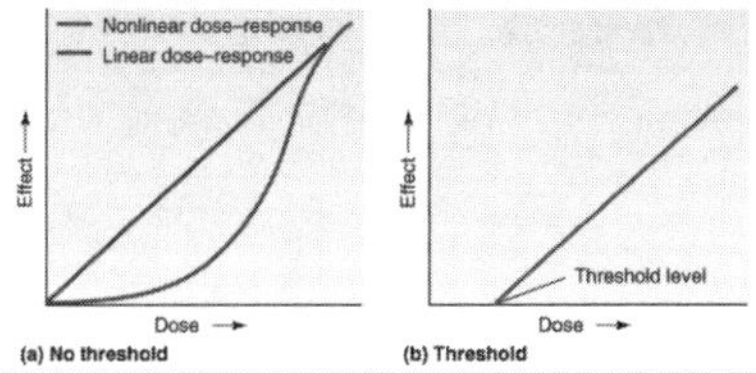

Nonlinear dose–response
Linear dose–response
Effect
Dose
(a) No threshold
Effect
Threshold level
Dose
(b) Threshold

CH18-32

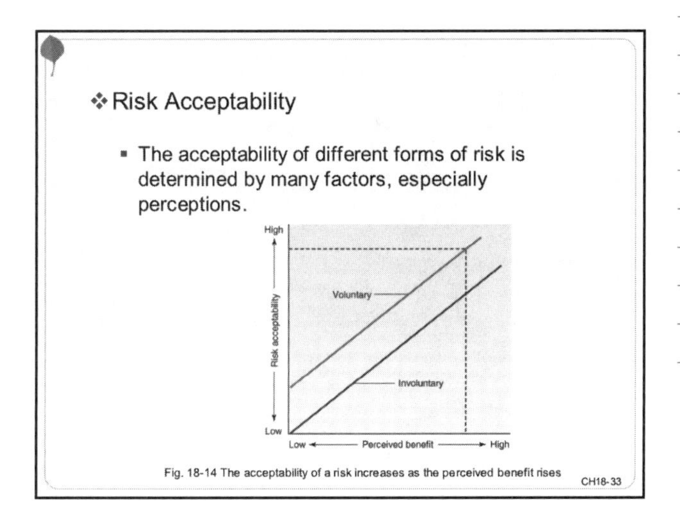

Fig. 18-14 The acceptability of a risk increases as the perceived benefit rises

CH18-33

Notes

Notes

❖ How Do We Decide If a Risk Is Acceptable?

- Several techniques are used to determine if the risk(s) posed by a technology or activity are acceptable.
- The most common is cost-benefit analysis, which weighs the costs against the benefits.
- Sustainable development strategies minimize social, economic, and environmental costs and maximize the benefits.

CH18-34

❖ How Do We Decide If a Risk Is Acceptable? (cont.)

- Risk assessment is designed to facilitate decision making by ensuring that the risk we perceive to be posed by any factor is equal to the actual risk.
- Our values, what we perceive as right or wrong, come from many sources and often affect our decisions.
- People's values tend to be somewhat narrow with respect to time and space.
- A long-range view that encompasses the entire planet is essential for sustainability.

CH18-35

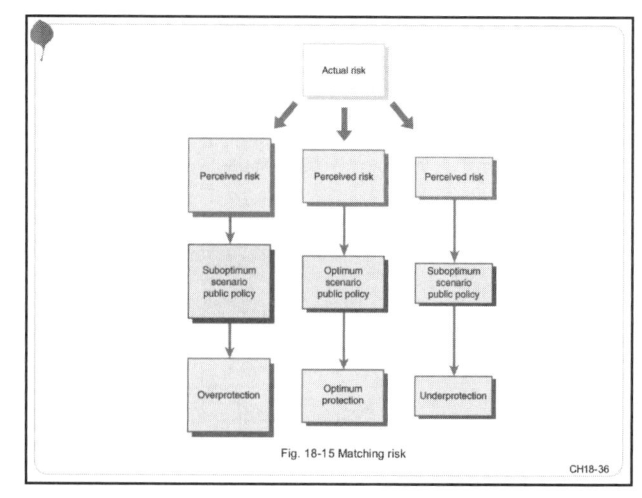

Fig. 18-15 Matching risk

CH18-36

Chapter 19: Air Pollution and Noise: Living and Working in a Healthy Environment

Notes

Environmental Science SEVENTH EDITION — Daniel D. Chiras

Chapter 19

Air Pollution and Noise

Living and Working in a Healthy Environment

CH19-1

19.1 Air: The Endangered Global Commons

- Air is a renewable resource cleansed by natural processes and regenerated by living things.
- This global resource is used by many and protected by few.
- It suffers from the tragedy that befalls many commons.

CH19-2

❖ Sources of Air Pollution

- Pollutants arise from natural and human or anthropogenic sources.
- Human-generated pollution is generally of greatest concern because it is produced in localized regions, so that concentrations reach potentially dangerous levels

Table 19-1
Natural Air Pollutants

Source	Pollutants
Volcanoes	Sulfur oxides, particulates
Forest fires	Carbon monoxide, carbon dioxide, nitrogen oxides, particulates
Wind storms	Dust
Plants (live)	Hydrocarbons, pollen
Plants (decaying)	Methane, hydrogen sulfide
Soil	Viruses, dust
Sea	Salt particulates

CH19-3

Notes

❖ Anthropogenic Air Pollutants and Their Sources

- Air pollutants come primarily from three sources:
  - Transportation
  - Energy production
  - Industry
- Combustion of fossil fuels is the major source of air pollution in these sectors.

CH19-4

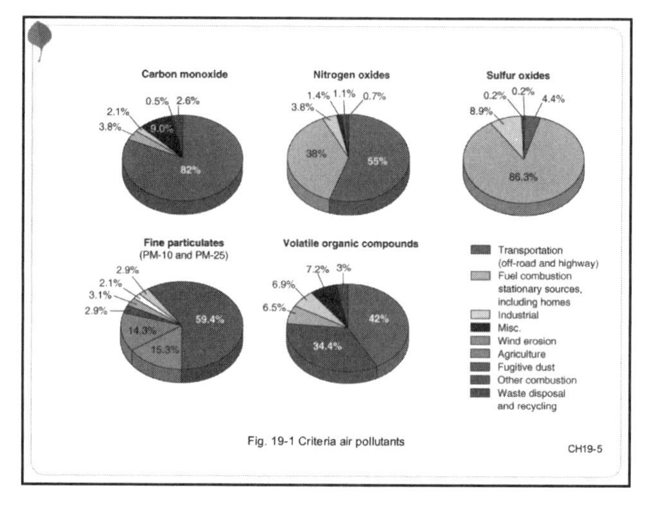

Fig. 19-1 Criteria air pollutants

CH19-5

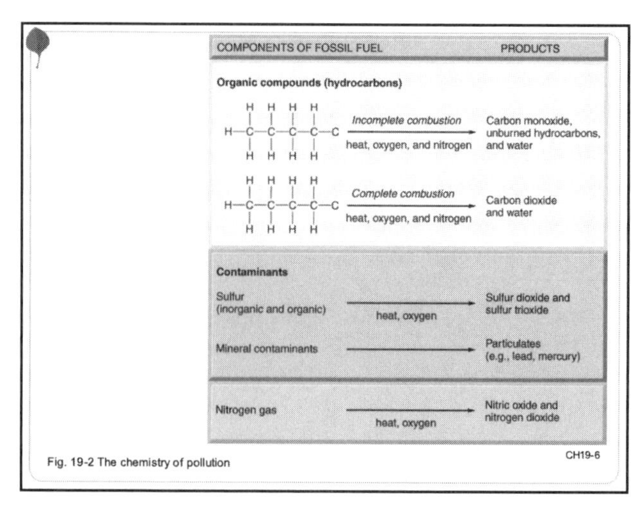

Fig. 19-2 The chemistry of pollution

CH19-6

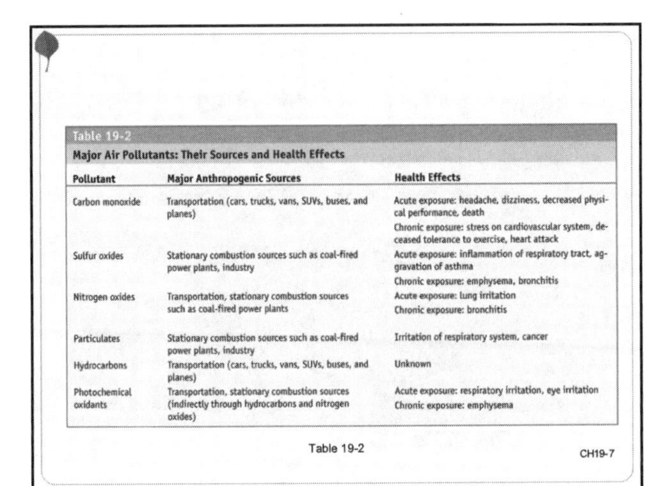

Table 19-2

Major Air Pollutants: Their Sources and Health Effects

Pollutant	Major Anthropogenic Sources	Health Effects
Carbon monoxide	Transportation (cars, trucks, vans, SUVs, buses, and planes)	Acute exposure: headache, dizziness, decreased physical performance, death Chronic exposure: stress on cardiovascular system, deceased tolerance to exercise, heart attack
Sulfur oxides	Stationary combustion sources such as coal-fired power plants, industry	Acute exposure: inflammation of respiratory tract, aggravation of asthma Chronic exposure: emphysema, bronchitis
Nitrogen oxides	Transportation, stationary combustion sources such as coal-fired power plants	Acute exposure: lung irritation Chronic exposure: bronchitis
Particulates	Stationary combustion sources such as coal-fired power plants, industry	Irritation of respiratory system, cancer
Hydrocarbons	Transportation (cars, trucks, vans, SUVs, buses, and planes)	Unknown
Photochemical oxidants	Transportation, stationary combustion sources (indirectly through hydrocarbons and nitrogen oxides)	Acute exposure: respiratory irritation, eye irritation Chronic exposure: emphysema

Table 19-2

CH19-7

Notes

❖ Primary and Secondary Pollutants

- Pollutants released into the atmosphere (primary pollutants) can react with one another.
- They can also react with atmospheric components, such as water vapor, to form secondary pollutants.

CH19-8

❖ Toxic Air Pollutants

- Much of the interest in the past three decades has focused on criteria air pollutants.
- Many other potentially toxic chemicals are released into the air each year from industry and other sources.

CH19-9

Notes

❖ Photochemical and Industrial Smog

- Smog is a term applied to two different types of air pollution.
- In cities with drier, sunnier climates and minimal industrial activity:
  - Hydrocarbons and nitrogen oxides from motor vehicles react to form a brownish haze called photochemical smog.

Fig. 19-3b Smog in Los Angeles

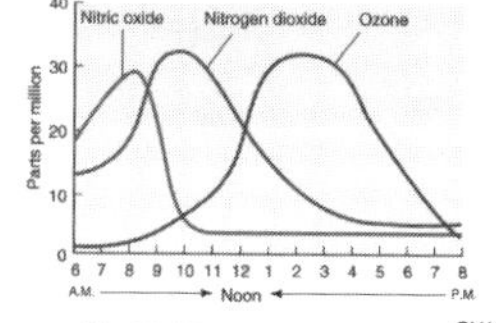

Fig. 19-4 Photochemical smog

CH19-10

❖ Photochemical and Industrial Smog (cont.)

- In older industrial cities with moister climates:
  - Particulates and sulfur oxides form industrial smog.

Fig. 19-3a Gray air smog

CH19-11

❖ Air Pollution – A Symptom of Unsustainable Systems?

- Air pollution, like other forms of environmental deterioration, is a symptom of unsustainable systems of:
  - Transportation
  - Industry
  - Housing
  - Energy production

CH19-12

19.2 The Effects of Climate and Topography on Air Pollution

- Air pollution levels in a region are affected by a number of factors, among them:
  - Temperature
  - Sunlight
  - Wind
  - Other climate factors
- They are also affected by the topography.

CH19-13

❖ The Cleansing Effects of Wind and Rain: Don't Be Fooled

- Wind and rain both tend to cleanse the air above our cities, but pollutants do not disappear.
- They're either blown elsewhere or fall from the sky, ending up in waterways or in our soil or on buildings and other structures.

CH19-14

❖ Mountains and Hills

- Mountains and hilly terrain can impede the flow of air, resulting in the buildup of pollutants in cities and industrialized areas.

CH19-15

Notes

Notes

❖ Temperature Inversions

- Temperature inversions result in the formation of warm-air lids that form over cities and even large regions.
- These lids impede the vertical mixing of air.
- This, in turn, traps cooler pollutant-laden air below.

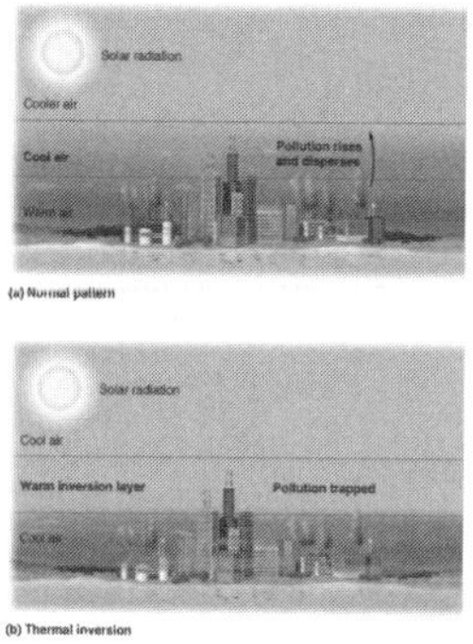

(a) Normal pattern

(b) Thermal inversion

Fig. 19-5 Temperature inversion

CH19-16

19.3 The Effects of Air Pollution

❖ Air pollution:

- adversely affects human health
- damages the environment and the organisms that live in it
- damages buildings and a wide assortment of materials, costing billions of dollars a year

CH19-17

❖ The Health Effects of Air Pollution

- Air pollution has a variety of health effects, ranging from immediate to delayed and from slight irritation to potentially life-threatening conditions.

CH19-18

- Air pollution causes many immediate effects, such as:
  - shortness of breath
  - eye irritation
  - upper respiratory tract irritation
- Few people are aware of the source of these problems.
- In extreme cases, pollutants can become lethal.

CH19-19

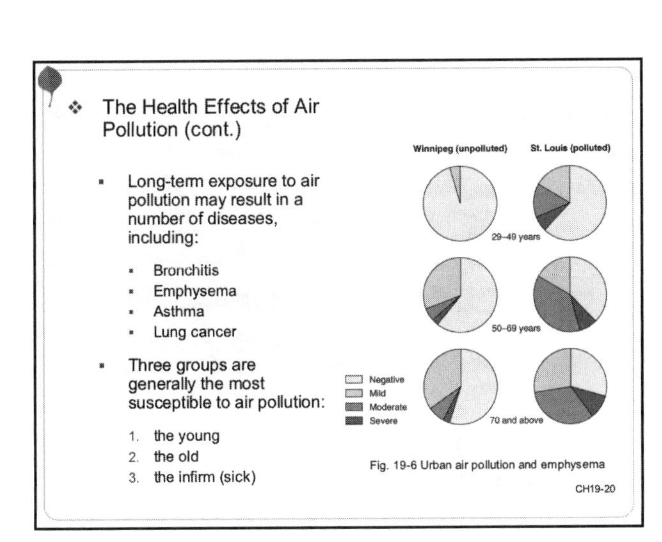

Fig. 19-6 Urban air pollution and emphysema

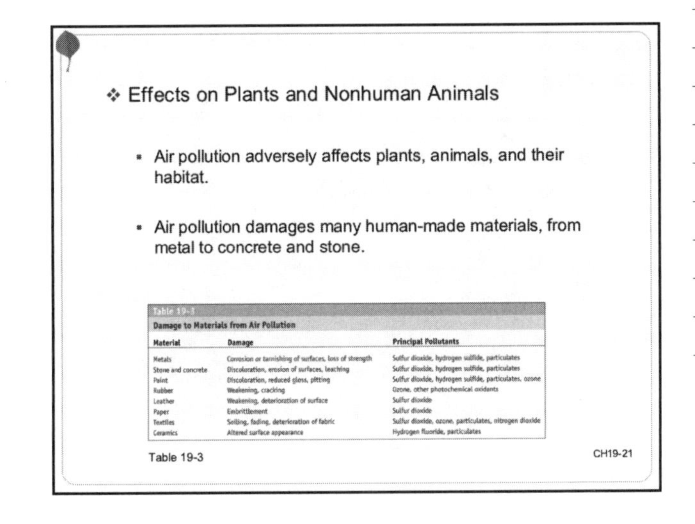

Table 19-3
Damage to Materials from Air Pollution

Material	Damage	Principal Pollutants
Metals	Corrosion or tarnishing of surfaces, loss of strength	Sulfur dioxide, hydrogen sulfide, particulates
Stone and concrete	Discoloration, erosion of surfaces, leaching	Sulfur dioxide, hydrogen sulfide, particulates
Paint	Discoloration, reduced gloss, pitting	Sulfur dioxide, hydrogen sulfide, particulates, ozone
Rubber	Weakening, cracking	Ozone, other photochemical oxidants
Leather	Weakening, deterioration of surface	Sulfur dioxide
Paper	Embrittlement	Sulfur dioxide
Textiles	Soiling, fading, deterioration of fabric	Sulfur dioxide, ozone, particulates, nitrogen dioxide
Ceramics	Altered surface appearance	Hydrogen fluoride, particulates

Table 19-3

Notes

Notes

19.4 Air Pollution Control: Toward a Sustainable Strategy

- Air pollution control is costly but appears to reap huge economic benefits – far in excess of the cost of controls.

CH19-22

- Cleaner Air through Better Laws
  - The Clean Air Act has been strengthened over the years to provide a comprehensive means of reducing air pollution.
  - It consists of many parts, including:
    - standards for emissions from various sources
    - air quality standards
    - several market-based incentives designed to reduce the emissions of pollutants

CH19-23

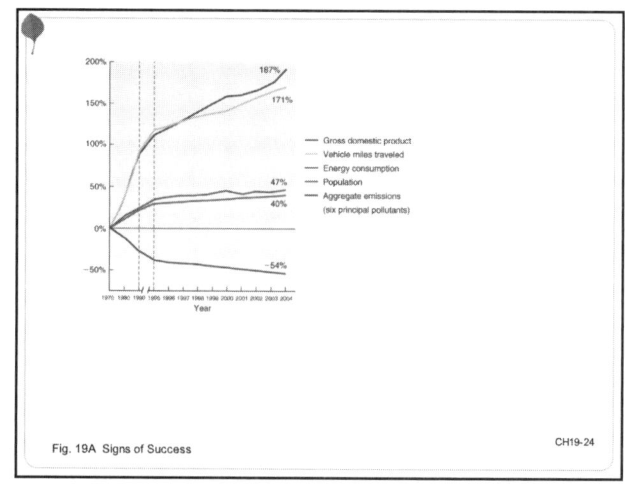

Fig. 19A Signs of Success

CH19-24

Notes

❖ Cleaner Air through Technology: End-of-Pipe Solutions

- To date, many efforts to control pollution have relied on end-of-pipe solutions.
- Most are pollution control devices that capture pollutants or convert them into (supposedly) less harmful substances.

CH19-25

❖ Cleaner Air through Technology: End-of-Pipe Solutions (cont.)

- Pollutants can be:
  - filtered from smokestack gases
  - precipitated out
  - washed out
- The problem with these techniques is that the pollutant must then be disposed of.
- Disposal in landfills can result in the pollution of groundwater.

CH19-26

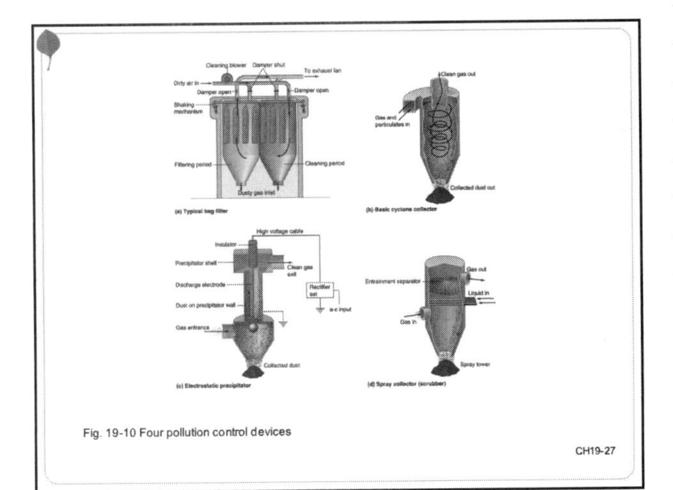

Fig. 19-10 Four pollution control devices

CH19-27

Notes

❖ Cleaner Air through Technology: End-of-Pipe Solutions (cont.)

- Emissions reductions from automobiles and other mobile sources are achieved by catalytic converters.
- These devices convert pollutants in the exhaust of cars to less harmful substances, at least from a human health perspective.

CH19-28

❖ New Combustion Technologies

- Burning coal and natural gas more efficiently in improved burners helps reduce the amount of pollution emitted per unit of energy consumption.
- This can help society meet energy demands more sustainably.

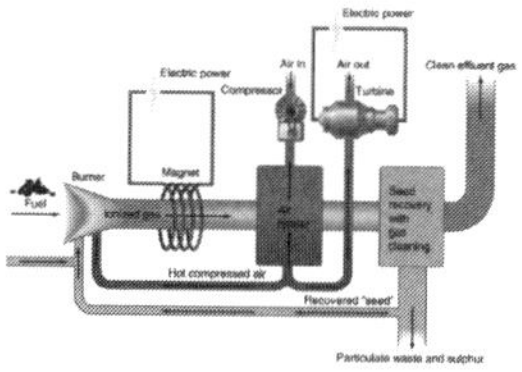

Fig. 19-12 Magnetohydrodynamics

CH19-29

❖ Economics and Air Pollution Control

- Air pollution control via end-of-pipe methods often adds substantial economic costs to various processes.
- It also produces waste products that, if not used for other purposes, end up in landfills.

CH19-30

❖ Toward a Sustainable Strategy

- Pollution prevention is a key element of a sustainable society.
- Pollution prevention results from enacting measures that promote:
  - Conservation (frugality and efficiency)
  - Recycling
  - Renewable resource use
  - Restoration of habitats

CH19-31

19.5 Noise: The Forgotten Pollutant

❖ What Is Sound?

- Sound waves are compression waves that travel through the air.
- Sound is characterized by loudness (measured in decibels) and pitch (how high or low it is).

CH19-32

❖ What Is Noise?

- Noise is an unwanted, unpleasant sound.
- What individuals consider to be a noise depends on many variables, such as the time of day or loudness of the sound.

CH19-33

Notes

Notes

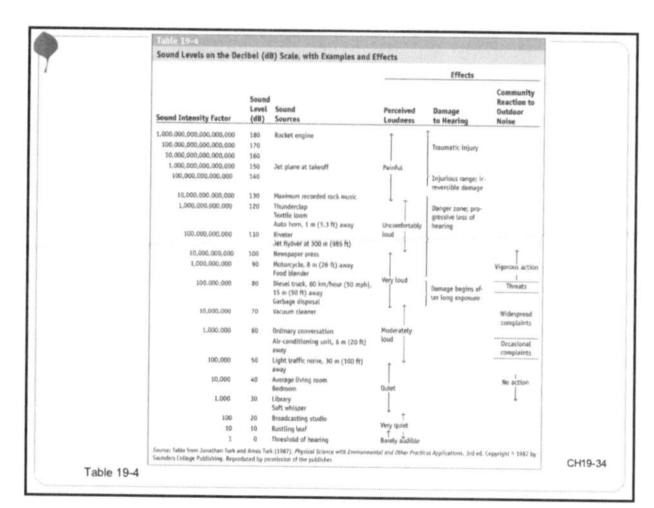

Table 19-4

Sound Levels on the Decibel (dB) Scale, with Examples and Effects

Sound Intensity Factor	Sound Level (dB)	Sound Sources	Effects: Perceived Loudness	Effects: Damage to Hearing	Effects: Community Reaction to Outdoor Noise
1,000,000,000,000,000,000	180	Rocket engine			
100,000,000,000,000,000	170			Traumatic injury	
10,000,000,000,000,000	160				
1,000,000,000,000,000	150	Jet plane at takeoff	Painful		
100,000,000,000,000	140			Injurious range; irreversible damage	
10,000,000,000,000	130	Maximum recorded rock music			
1,000,000,000,000	120	Thunderclap Textile loom Auto horn, 1 m (3.3 ft) away		Danger zone; progressive loss of hearing	
100,000,000,000	110	Riveter Jet flyover at 300 m (985 ft)	Uncomfortably loud		
10,000,000,000	100	Newspaper press			
1,000,000,000	90	Motorcycle, 8 m (26 ft) away Food blender			Vigorous action
100,000,000	80	Diesel truck, 80 km/hour (50 mph), 15 m (50 ft) away Garbage disposal	Very loud	Damage begins after long exposure	Threats
10,000,000	70	Vacuum cleaner			Widespread complaints
1,000,000	60	Ordinary conversation Air-conditioning unit, 6 m (20 ft) away	Moderately loud		Occasional complaints
100,000	50	Light traffic noise, 30 m (100 ft) away			
10,000	40	Average living room Bedroom	Quiet		No action
1,000	30	Library Soft whisper			
100	20	Broadcasting studio			
10	10	Rustling leaf	Very quiet		
1	0	Threshold of hearing	Barely audible		

Source: Table from Jonathan Turk and Amos Turk (1987). *Physical Science with Environmental and Other Practical Applications*, 3rd ed. Copyright © 1987 by Saunders College Publishing. Reproduced by permission of the publisher.

CH19-34

❖ Impacts of Noise

- Noise affects us in many ways. It:
  - damages hearing
  - disrupts our sleep
  - annoys us in our everyday lives
- It interferes with conversation, concentration, relaxation, and leisure.

CH19-35

❖ Controlling Noise

- Noise levels can be controlled by:
  - redesigning machinery and other sources
  - sound-insulating buildings
  - separating noise generators from people
  - other measures

CH19-36

19.6 Indoor Air Pollution

- Indoor air pollutants come from:
  - combustion sources
  - a variety of products that release potentially harmful substances
  - from naturally occurring radioactive materials in the ground beneath a building

CH19-37

Notes

❖ How Serious Is Indoor Air Pollution?

- Indoor air pollution is present in millions of homes and offices and adversely affects millions of Americans.

Table 19-6
Major Indoor Air Pollutants

Pollutant	Source(s)	Effects on Human Health
Tobacco smoke—carbon monoxide, particulates, sulfur dioxide, nitrogen dioxide	Cigarettes, pipes, and cigars	Carbon monoxide—shortness of breath and heart attacks; particulates—lung cancer; sulfur and nitrogen dioxide—emphysema and chronic bronchitis. Effects seen in smokers and nonsmokers.
Radon	Naturally occurring radioactive material (radium) in the soil. Radon leaks through cracks in the foundation	Lung cancer
Formaldehyde	New furniture, particleboard, plywood, chipboard, and wood paneling	Irritation of the eyes, nose, and throat; nasal cancer and lung cancer
Asbestos	Insulation around pipes in old homes, school buildings, and factories	Lung cancer and buildup of fibrous tissue in the lungs, which makes breathing difficult

CH19-38

❖ Why Is Indoor Air in Buildings So Polluted?

- Indoor air pollution has been around for centuries.
- Today, problems are created by:
  - efforts to reduce air infiltration which traps pollutants
  - the use of many new building materials and products that contain glues and other chemicals that are harmful to human health

CH19-39

Chapter 20: Global Air Pollution: Ozone Depletion, Acid Deposition, and Global Warming

Notes

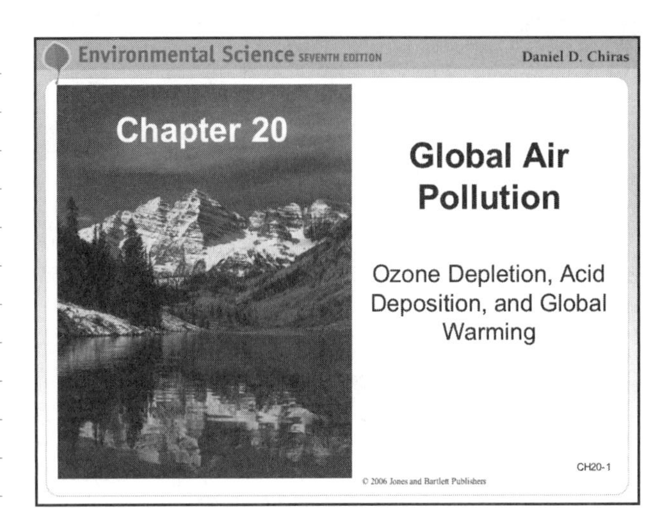

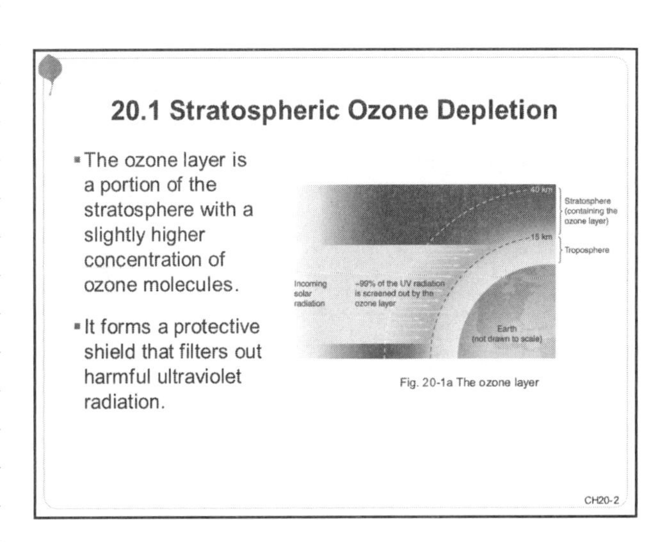

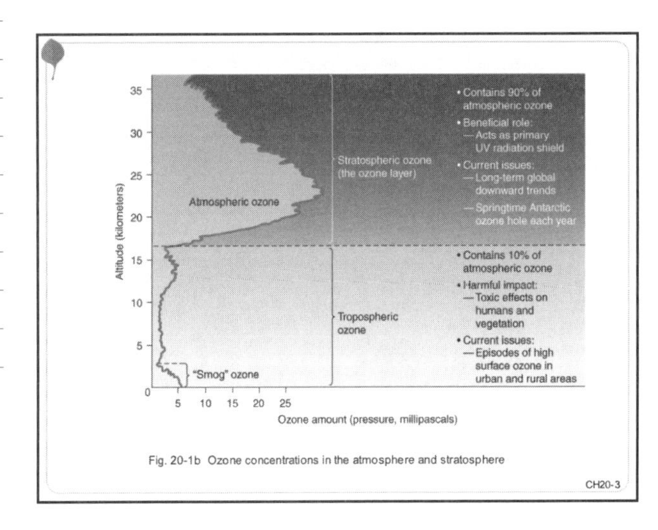

Notes

❖ Activities That Deplete the Ozone Layer

- Human civilization threatens the ozone layer through two principal activities:

  1. the use of a class of chemical compounds called chlorofluorocarbons (CFCs)
  2. jet travel through the stratosphere

CH20-4

❖ Activities That Deplete the Ozone Layer (cont.)

- Chlorofluorocarbon molecules were once used as:
  - Propellants
  - Refrigerants
  - Blowing agents
  - Cleaning agents

CH20-5

❖Activities That Deplete the Ozone Layer (cont.)

- CFCs are stable molecules that diffuse into the stratosphere, where they break down, releasing chlorine atoms.
- Chlorine atoms react with ozone molecules, destroying them.
- Other chlorine and bromine-containing compounds have also been used widely and are known as ozone depleters.

Fluorine
Carbon
Chlorine
UV radiation
Chlorofluorocarbon (CFC) molecule
(a)
Ultraviolet radiation from the sun strikes the CFC molecule and causes a chlorine atom to break away.
Chlorine free radical
Ozone molecule made up of three oxygen atoms (O_3)
Chlorine oxide
Oxygen (O_2)
(b)
The chlorine atom reacts with an ozone molecule to form chlorine oxide and diatomic oxygen.
Oxygen atom
Chlorine oxide
Oxygen (O_2)
Chlorine free radical
(c)
When a free atom of oxygen reacts with a chlorine oxide molecule, diatomic oxygen is formed, and the chlorine atom is released to destroy more ozone.

Fig. 20-2 The chemistry of CFCs and ozone depletion.

CH20-6

❖ Activities That Deplete the Ozone Layer (cont.)

- All jets release nitric oxide gas, a pollutant that reacts with ozone.
- Jets that travel through the stratosphere, such as the SST, however, have the greatest impact.
- Because few high-flying jets are in use today other than in the military, jet travel poses a lower risk than the use of CFCs.

CH20-7

❖ Ozone Depletion: The History of a Scientific Discovery

- Studies of the ozone layer show substantial declines over the globe, with the highest level of depletion in the southern hemisphere and Antarctica.

CH20-8

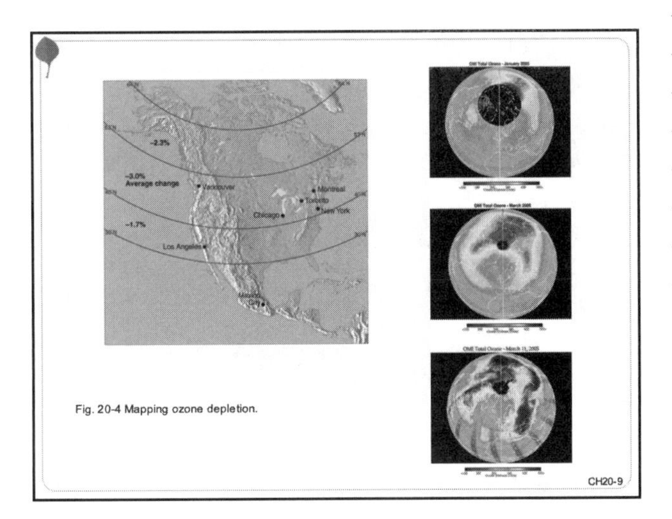

Fig. 20-4 Mapping ozone depletion.

CH20-9

Notes

Notes

❖ The Many Effects of Ozone Depletion

- Ozone depletion is resulting in an increase in ultraviolet radiation striking the Earth, especially in unpolluted areas.
- Ultraviolet radiation causes:
  - **Skin cancer**
  - **Cataracts**
  - **Premature aging**
- It could also seriously damage:
  - **Ecosystems**
  - **Crops**
  - **Materials**
  - **Finishes**

CH20-10

❖ Banning Ozone-Depleting Chemicals: A Global Success Story

- As scientific evidence on ozone depletion accumulated, world nations tightened restrictions on the production of ozone-depleting chemicals.
- Three international treaties have already been signed to eliminate the production of ozone-depleting compounds.
- Progress has been made toward meeting these goals.

Fig. 20-5 Graphing success

CH20-11

❖ Substitutes for Ozone-Destroying CFCs

- Most CFCs have been being replaced by a class of compounds called HCFCs.
- These less stable compounds deplete the ozone layer, but not as rapidly as their predecessors.
- Because of this, they are viewed mainly as an interim solution.
- Efforts are under way to replace CFCs used for cleaning agents with more friendly water-soluble agents.

CH20-12

Notes

❖ The Good News and Bad News about Ozone

- The accumulation of CFCs and other ozone-depleting compounds in the atmosphere has begun to slow.
- Despite this progress, the ozone layer will take many years to recover.
- Many people will develop and die from skin cancer, but the phase-out will also spare many people as well.

CH20-13

20.2 Acid Deposition

- Acid deposition from pollutants is a global problem with serious social, economic, and environmental impacts.

CH20-14

❖ What Is an Acid?

- Acids are chemical substances that add hydrogen ions to a solution.
- Acids are measured on the pH scale, which ranges from 0 to 14, with 7 being neutral—neither acidic nor basic.

Lemon juice
Vinegar
Mean pH of Adirondack Lakes-1975
"Pure" rain (5.7)
Mean pH of Adirondack Lakes-1930s
Distilled water
Baking soda
Acid rain
0 1 2 3 4 5 6 7 8 9 10 11 12 13 14
Acidic ← Neutral → Basic

Fig. 20-7 the pH scale

CH20-15

Notes

❖ What Is Acid Deposition?

- Rainfall in unpolluted areas has a pH of about 5.7 and is just slightly acidic.
- Acid deposition refers to rain and snow with a pH of less than 5.7 and the deposition of acid particles and gases.
- Acids reach the surface of the Earth either as wet deposition (rain or snow) or dry deposition (particulates and gases).

CH20-16

❖ Where Do Acids Come From?

- Acid precursors come from natural and anthropogenic sources, the latter being the most important.
- Of the anthropogenic sources, the combustion of fossil fuels is the most significant.

Fig. 20-8 Sources of sulfur dioxide

CH20-17

❖ The Transport of Acid Precursors

- Acid precursors can be transported hundreds of kilometers from their site of production to their site of deposition.
- Acid deposition occurs downwind from virtually all major industrial and urban centers.
- Acid deposition is increasing in strength (acidity) and expanding in geographic range.

CH20-18

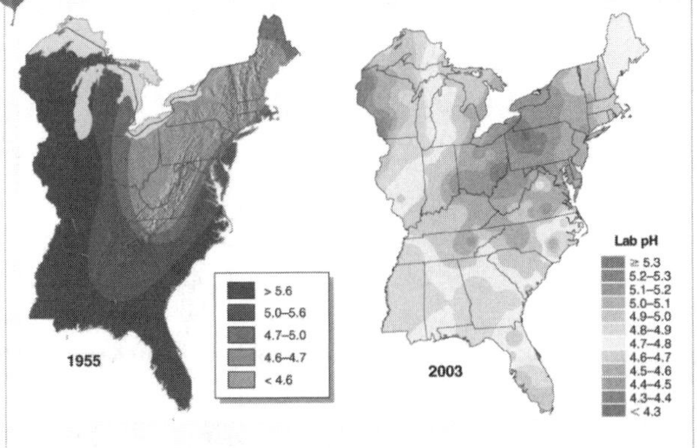

Notes

1955

2003

> 5.6
5.0–5.6
4.7–5.0
4.6–4.7
< 4.6

Lab pH
≥ 5.3
5.2–5.3
5.1–5.2
5.0–5.1
4.9–5.0
4.8–4.9
4.7–4.8
4.6–4.7
4.5–4.6
4.4–4.5
4.3–4.4
< 4.3

Fig. 20-9 Acid precipitation has increased in the eastern U.S. between 1955 and 2003.

CH20-19

❖ The Social, Economic, and Environmental Impacts of Acid Deposition

- Acid deposition has acidified lakes throughout the world.
- Hundreds of lakes no longer support aquatic life, and thousands are on the verge of ecological collapse.

CH20-20

❖ The Social, Economic, and Environmental Impacts of Acid Deposition (cont.)

- Soil and surface waters have a buffering capacity – an ability to resist changes in pH.
- The buffering capacity plays an important role in determining if a lake will be damaged by acid deposition.

CH20-21

Notes

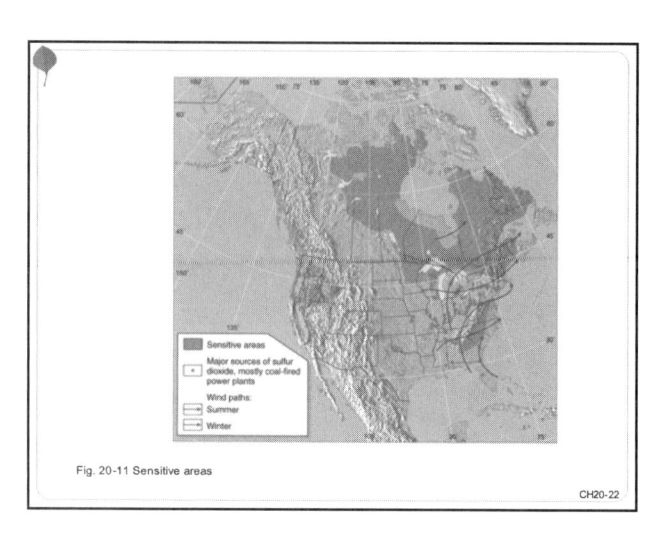

Fig. 20-11 Sensitive areas

CH20-22

- The Social, Economic, and Environmental Impacts of Acid Deposition (cont.)
  - Acidity kills aquatic organisms and impairs growth and reproduction.
  - Acidity also leaches heavy metals, toxic to fish, from the soil.

CH20-23

- The Social, Economic, and Environmental Impacts of Acid Deposition (cont.)
  - Acid deposition affects birds living near lakes and aquatic species such as salamanders, which are a key element of the food chain.

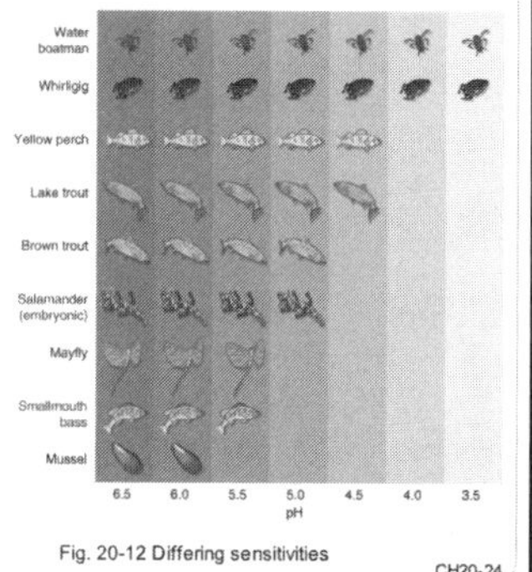

Fig. 20-12 Differing sensitivities

CH20-24

❖ The Social, Economic, and Environmental Impacts of Acid Deposition (cont.)

- Acid deposition damages forests in many parts of the world and may affect crops as well.
- Trees and other plants are damaged directly by acids.
- They are also damaged indirectly through changes in the soil chemistry and soil-dwelling organisms.

Fig. 20-14 Forest die-off

CH20-25

Notes

❖ The Social, Economic, and Environmental Impacts of Acid Deposition (cont.)

- The sulfur and nitrogen in sulfuric and nitric acid promote plant growth.
- Their negative effects (such as direct damage and changes in the soil chemistry) typically outweigh any benefits resulting from their fertilizing effect.

❖ Damage to Materials

- Acids cause billions of dollars of damage to priceless statues, buildings, and materials.

CH20-26

❖ Solving a Growing Problem – Short-Term Solutions

- Many stopgap measures have been initiated to help reduce the threat of acid deposition, including:
  - installation of smokestack scrubbers
  - combustion of low-sulfur or desulfurized coal
  - liming lakes to neutralize acidity

CH20-27

Notes

❖ Long-Term Sustainable Strategies

- A sustainable design strategy to help prevent the production of acid precursors (and thus reduce acid deposition) includes:
  - fuel efficiency
  - renewable fuels
  - recycling
  - population stabilization
  - growth management

CH20-28

❖ Are Controls on Acid Deposition Working?

- Market-based strategies, such as tradable permits, have proven successful in reducing sulfur dioxide emissions in the U.S.
- Corresponding changes are being seen in the acidity of rainfall as well as many lakes and streams.

CH20-29

20.3 Global Warming/Global Climate Change

❖ Global Energy Balance and the Greenhouse Effect

- Much of the sunlight striking the Earth and its atmosphere is converted into heat and is radiated back into space.
- Natural and anthropogenic factors affect the amount of solar radiation striking the Earth and the rate at which heat escapes.
- Therefore, these factors influence the temperature of the Earth's atmosphere.

CH20-30

Total reflected 32%

21% Reflected from clouds

5% Reflected from dust

1-2% Absorbed by plants

67% Heat

Absorbed by air, land and water and converted to heat

6% Reflected from water, soil, air, vegetation

(a)

Water vapor

CH_4

NO_2

CO_2

Heat

Heat

Heat

Heat

(b)

Fig. 20-17 Global energy balance

CH20-31

❖ Greenhouse Gases: Where Do they Come From?

- Greenhouse gases come from natural and anthropogenic sources.
- Anthropogenic sources have been increasing dramatically over the past 50 years.

Table 20-2

Major Greenhouse Gases and Their Characteristics

Gas	Atmospheric Concentration (ppm)	Annual Increase (%)	Life Span (Years)	Relative Greenhouse Efficiency (CO_2 = 1)	Current Greenhouse Contribution (%)	Principal Sources of Gas
Carbon dioxide (CO_2) (from fossil fuels)	351.3	0.4	x[1]	1	57 (44)	Coal, oil, natural gas, deforestation
Carbon dioxide (from biological sources)					(13)	
Chlorofluorocarbons (CFCs)	0.000225	5	75–111	15,000	25	Foams, aerosols, refrigerants, solvents
Methane (CH_4)	1.675	1	11	25	12	Wetlands, rice, fossil fuels, livestock
Nitrous oxide (N_2O)	0.31	0.2	150	230	6	Fossil fuels, fertilizers, deforestation

Sources: Data from World Watch Institute, U.S. EPA, and *Journal of Geophysical Research*.

[1]Carbon dioxide is a stable molecule with a 2- to 4-year average residence time in the atmosphere.

CH20-32

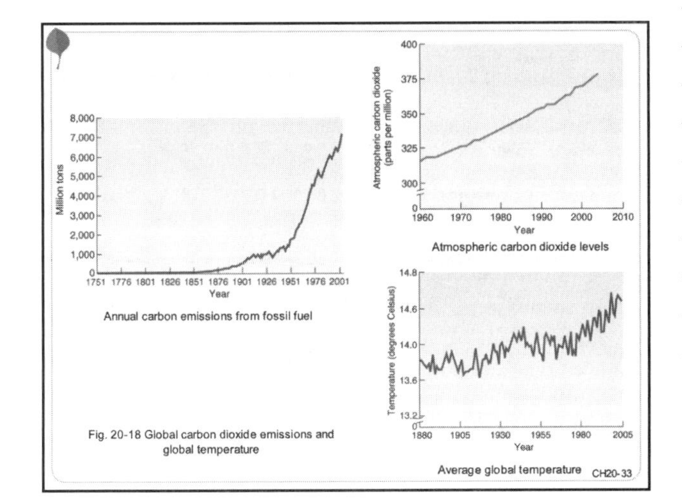

Fig. 20-18 Global carbon dioxide emissions and global temperature

CH20-33

Notes

Notes

❖ Upsetting the Balance: Global Warming and Global Climate Change

- The accumulation of greenhouse gases is very likely responsible for the increase in average daily temperatures.
- This is linked to changes in other aspects of climate, such as:
  - rainfall patterns
  - the frequency and severity of storms
- Many of these changes affect ecosystems.

CH20-34

❖ Predicting the Effects of Greenhouse Gases

- Scientists predict the effect of greenhouse gases by using computer programs that simulate global climate.
- They use information on projected levels of greenhouse gases to determine future temperature and other climatic effects.

CH20-35

❖ Predicting the Effects of Greenhouse Gases (cont.)

- A small but climatically significant increase in global temperature is expected by the end of the next century.
- This increase is a result of increasing emission of greenhouse gases.
- Scientists predict that this increase will very likely result in a rise in sea level with potentially devastating effects on coastal human population centers.

Fig. 20-19 Paradise lost

CH20-36

❖ Predicting the Effects of Greenhouse Gases (cont.)

- Global temperature increases could shift rainfall patterns, increasing precipitation in some areas and decreasing it in others.

- Too little rain in some areas and too much in others could affect food production and economic output.

CH20-37

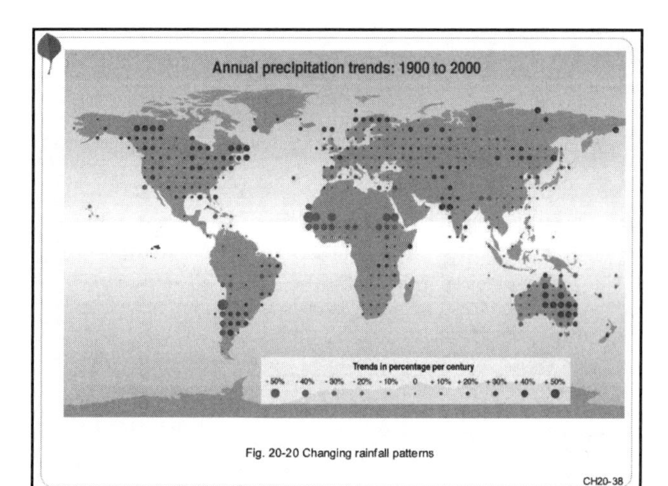

Fig. 20-20 Changing rainfall patterns

CH20-38

❖ Predicting the Effects of Greenhouse Gases (cont.)

- Preliminary studies suggest that global warming may be responsible for an increase in the number and severity of storms across the globe.

* Some climatologists are reluctant to pin any blame on global warming, but the 2005 hurricane season shattered records that have stood for decades.
  - Named storms: 27; previous record: 21 in 1933
  - Hurricanes: 13; previous record: 12 in 1969
  - Major hurricanes hitting the U.S.: Four (Dennis, Katrina, Rita and Wilma); previous record: Three, most recently in 2004
  - Hurricanes of Category 5 intensity (greater than 155 mph): Three (Katrina, Rita and Wilma); previous record: Two in 1960 and 1961

CH20-39

Notes

Notes

❖ The Ecological and Health Impacts of Global Climate Change

- Organisms and ecosystems could be profoundly influenced by global climate change.
- The influence will be especially strong if the rate of change occurs faster than their ability to adapt (which seems inevitable).

CH20-40

❖ Changing Ocean Current

- Sea water circulates throughout the oceans
- Warm water from the tropics flow north in the Atlantic and help warm northern landmasses.
- A massive melting of glaciers could add enough cool, fresh water to the oceans to disrupt the global currents.

CH20-41

❖ National and International Security

- Global climate change could result in civil unrest, including war, as countries try to defend and secure dwindling food, water, and energy supplies.

CH20-42

Notes

❖ Is Global Climate Change Occurring?

- Evidence shows many indications that global climate change is occurring.
  - Global carbon dioxide levels are increasing
  - Sea level is on the rise
  - Polar ice is diminishing
  - Global temperatures are rising
  - The global conveyor belt of ocean currents could be slowing down

CH20-43

❖ Uncertainties: What We Don't Know

- Factors that may result in a rapid increase in global carbon dioxide levels, causing accelerated planetary warming include:
  - warming of the oceans
  - melting of land-based ice
  - loss of forests
- Other factors, however, may offset these changes.

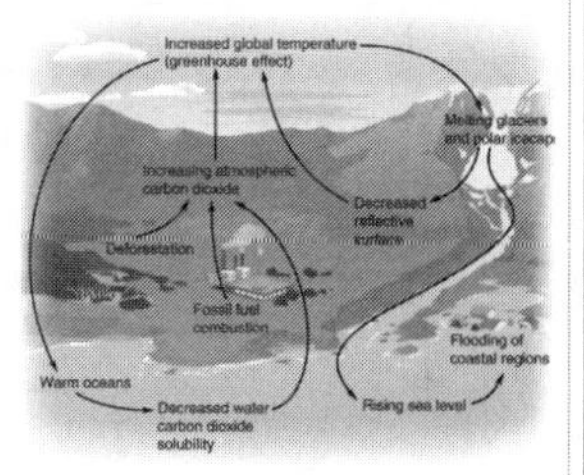

Fig. 20-23 Dangerous feedback

CH20-44

❖ Solving a Problem in a Climate of Uncertainty: Weighing Risks and Benefits

- Uncertainty exists on the global climate change issue, which has slowed progress toward solutions.
- Many believe that the costs of reducing or eliminating greenhouse gases are outweighed by the potential costs of global climate change.

CH20-45

Notes

❖ Solving the Problem Sustainably

- Redesigning human systems according to sound principles of sustainability could help alleviate the problem of global warming.
- Population stabilization can help reduce our demand for fossil fuels and other greenhouse-enhancing activities such as deforestation.
- Restoring forests, especially in the tropics, could have a profound effect on global carbon dioxide levels.

CH20-46

❖ Solving the Problem Sustainably (cont.)

- Recycling and energy efficiency greatly reduce energy demand and cut greenhouse gas emissions.

Table 20-3

Energy Consumption per Use for 12-Ounce Beverage Containers

Container	Energy Use (BTUs)
Aluminum can, used once	7050
Steel can, used once	5950
Recycled steel can	3880
Glass beer bottle, used once	3730
Recycled aluminum can	2550
Recycled glass beer bottle	2530
Refillable glass bottle, used 10 times	610

Source: From L.R. Brown, C. Flavin, and S. Postel (1991). *Saving the Planet: How to Shape an Environmentally Sustainable Economy*. New York: Norton.

CH20-47

❖ Solving the Problem Sustainably (cont.)

- Renewable energy technologies can provide us with much-needed power, with little or no impact on global climate.
- Solving the problem of global warming will require the efforts of all sectors of society and every country on Earth.

CH20-48

Chapter 21: Water Pollution: Sustainably Managing a Renewable Resource

Notes

Chapter 21

Water Pollution

Sustainably Managing a Renewable Resource

CH21-1

21.1 Surface Water Pollution

- Nations produce many different types of water pollutants.
- The water in all countries is plagued with pollutants from human and animal wastes.
- In industrial nations toxic chemicals also contribute to water pollution.

CH21-2

❖ Sources of Water Pollution

- Water pollutants arise from natural and anthropogenic sources.
- They travel freely from one location to another through rivers, streams, and groundwater.
- They are also transported from one medium (land or air) to another (water).

CH21-3

Notes

❖ Point and Nonpoint Sources

- Water pollution arises from identifiable **point** sources, such as factories, and from diffuse **nonpoint** sources, such as farm fields and streets.
- Point sources are much easier to control.

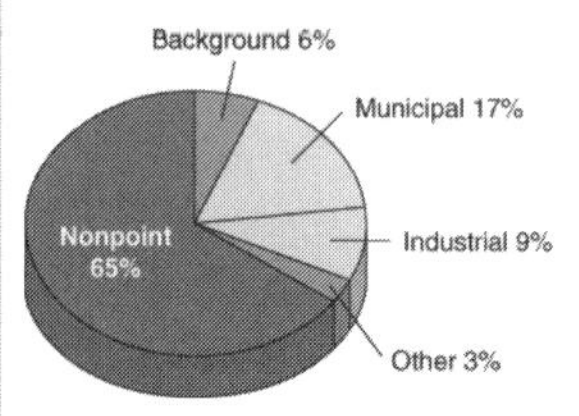

Fig. 21-2 Major sources of U.S. stream pollution

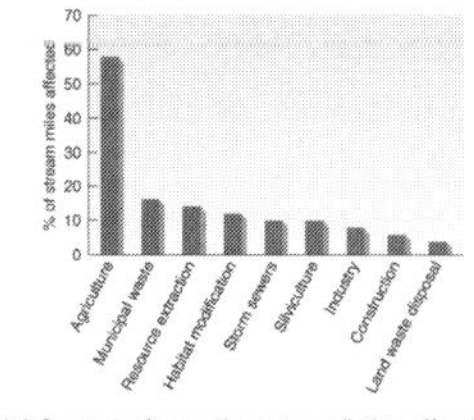

Fig. 21-3 Sources of nonpoint water pollution affecting streams

CH21-4

❖ Organic Nutrients

- Organic nutrients come from a variety of sources, primarily treated and untreated waste released into waterways.
- Organic compounds stimulate bacterial growth, which depletes oxygen levels in water bodies and kills off oxygen-dependent species.
- Oxygen levels can return to normal levels only if the influx of organic materials ceases.

CH21-5

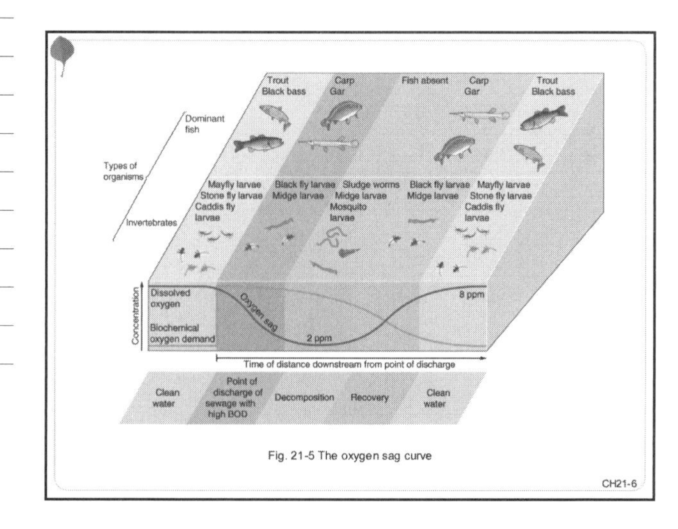

Fig. 21-5 The oxygen sag curve

CH21-6

❖ Inorganic Nutrients—Nitrates and Phosphates

- Inorganic nutrients stimulate excess plant growth which disrupts the aquatic environment.
- When plants die and decompose, oxygen levels decline sharply which can be harmful to many organisms.

CH21-7

❖ Eutrophication and Natural Succession

- The accumulation of nutrients in lakes is called natural or cultural **eutrophication**.
- Combined with sediment deposited from human activities, cultural eutrophication causes lakes to age prematurely.

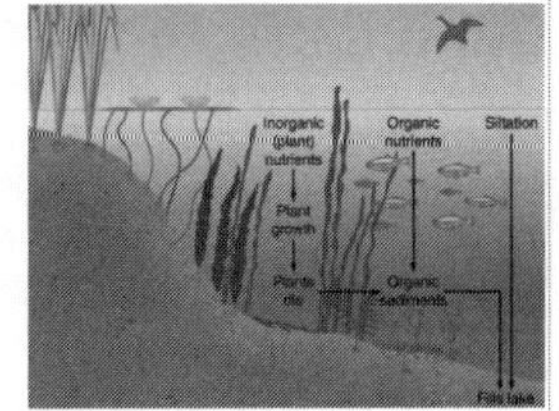

Fig. 21-6 Eutrophication and succession

CH21-8

❖ Infectious Agents

- Numerous infectious agents are found in surface waters.
- Sewage treatment facilities and drinking water purification have greatly reduced the incidence of disease in more developed countries, although outbreaks do occur.
- To monitor infectious agents, officials measure fecal coliform levels, a harmless bacterium itself, but an indicator of the presence of fecal contamination.

CH21-9

Notes

Notes

❖ Toxic Organic Water Pollutants

- Numerous toxic chemicals enter the waterways from factories, homes, farms, lawns, and gardens.
- Numerous inorganic pollutants such as acids and heavy metals make their way into the surface and groundwater of industrial nations, usually from industrial sources.
- Some, such as lead and mercury, are of major concern.

CH21-10

- Mercury is emitted from many sources and is one of the most common and most toxic inorganic pollutants, primarily exerting its effect through the nervous system.
- Nitrates can be converted to nitrites, which bind to hemoglobin and reduce the oxygen-carrying capacity of the blood.
- Road salts used to remove ice and make driving safe can profoundly affect aquatic ecosystems, forests, and orchards.
- Chlorine is added to water to kill harmful organisms. However, it also reacts with organic compounds to form substances toxic to humans and other organisms.

CH21-11

❖ Sediment

- Sediment washed from the land has profound effects on chemical and physical nature of ecosystems.
- Such changes have large impacts on aquatic organisms and humans who depend on them.

CH21-12

❖ Thermal Pollution

- Water from rivers and lakes is used to cool many industrial processes.
- Electric power production is one of the major processes.
- Heat generated by these processes is often discharged directly into surface waters, where it kills organisms or shifts the composition of the aquatic system.

CH21-13

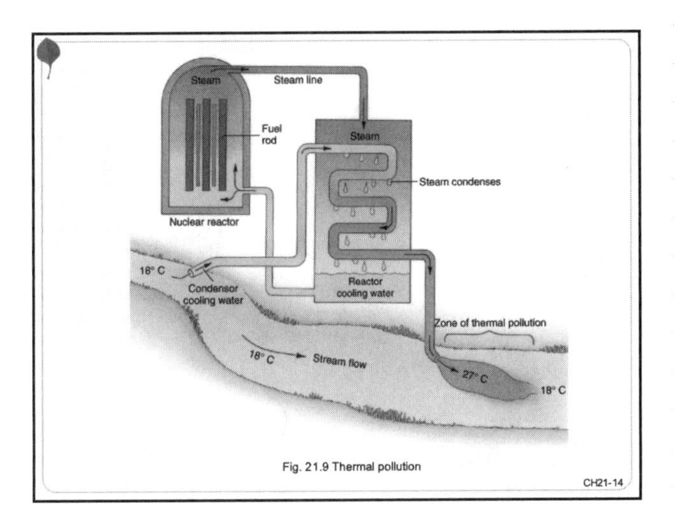

Fig. 21.9 Thermal pollution

CH21-14

21.2 Groundwater Pollution

- Groundwater is an important source of drinking water.
- It may be heavily contaminated in numerous industrialized nations by:
  - industrial waste pits
  - septic tanks
  - oil wells
  - landfills
- Groundwater in some rural areas may also be contaminated by agricultural chemicals, notably pesticides and fertilizer.

CH21-15

Notes

Notes

❖ Effects of Groundwater Pollution

- Thousands of chemicals may be found in a nation's groundwater.
- Many of them are potentially harmful to human health, causing problems for:
  - unborn children:
    - miscarriage
    - birth defects
    - premature infant death
  - adults:
    - rashes
    - neurological problems

CH21-16

❖ Cleaning Up Groundwater

- Groundwater moves slowly and takes many years to cleanse itself.
- Preventing groundwater pollution is essential to creating a sustainable water supply.
- Equally important are efforts to clean up groundwater supplies already contaminated by potentially toxic chemicals.

CH21-17

21.3 Ocean Pollution

- The oceans are polluted by:
  - chemicals spilled into them directly
  - pollutants washed from the lands and transported to them by rivers

CH21-18

❖ Oil in the Seas

- Half the oil polluting the oceans comes from natural seepage.
- The rest comes from human sources including tanker accidents and inland disposal.

CH21-19

❖ Oil in the Seas (cont.)

- Oil spilled from human sources:
  - evaporates
  - is broken down by naturally occurring bacteria
  - sinks to the bottom
- Before it is eliminated, however, it can cause serious environmental damage.
- The extent of the damage depends on:
  - the amount spilled
  - the location of the spill
  - the prevailing weather conditions
  - the season

CH21-20

❖ Oil in the Seas (cont.)

- Many efforts are under way in the United States and other countries to reduce oil spills, including new standards for the construction of oil tankers.
- Conservation, recycling, and renewable resource use can all reduce our dependence on oil and reduce oil pollution in the seas.

CH21-21

Notes

Notes

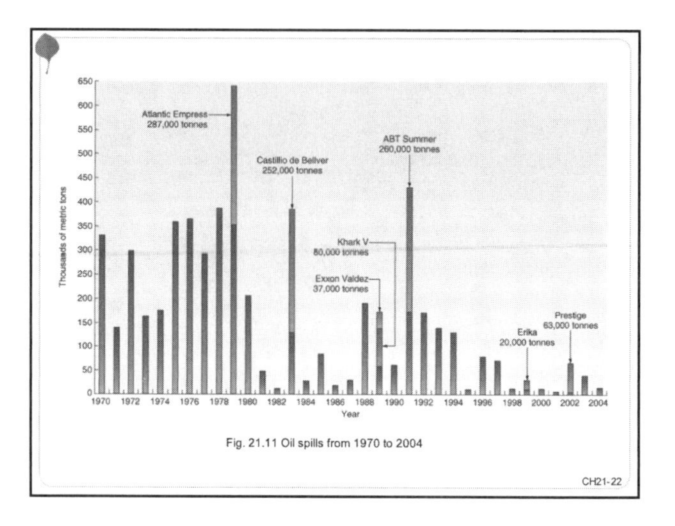

Fig. 21.11 Oil spills from 1970 to 2004

CH21-22

❖ Plastic Pollution

- Millions of tons of plastic are dumped into the ocean each year, killing hundreds of thousands of marine mammals, fish, and birds.
- Many steps have been taken to reduce the disposal of plastic into the ocean, but huge amounts are still being disposed of each year.

Fig. 21-13 Trash on a California beach

CH21-23

❖ Medical Wastes and Sewage Sludge

- Medical wastes and sewage have long been dumped into the ocean.
- Steps have been taken to eliminate both practices.
- The direct dumping of sewage has stopped entirely, but millions of gallons of sewage enter the sea each year from coastal sewage treatment plants.

CH21-24

❖ Red Tide

- Outbreaks of microscopic and often highly toxic algae appear to be on the rise worldwide.
- This may be caused by an increase in inorganic nutrient pollution from agriculture, industry, and the human population.

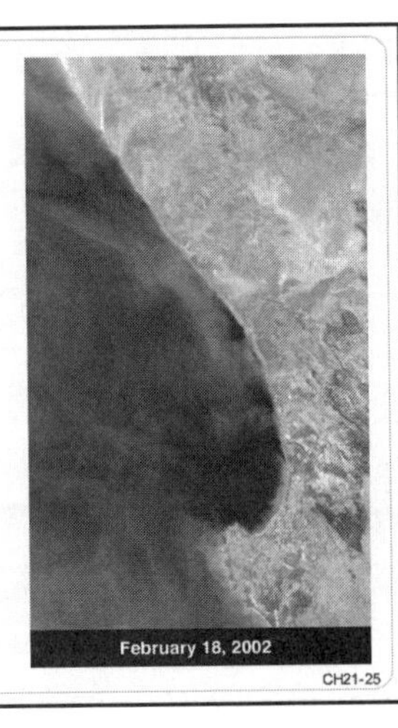

Red tide off coast of South Africa
Image courtesy NASA/GSFC/LaRC/JPL, MISR Team.

CH21-25

❖ The Case of the Dying Seals

- A massive seal die-off in the late 1980s, caused by a virus, may ultimately have resulted from immune system suppression caused by a common pollutant, PCBs.

Harbor seal
Courtesy of NOAA

CH21-26

21.4 Water Pollution Control

❖ Reducing water pollution requires efforts on two levels:

- those that capture wastes emitted from various sources (the so-called end-of-pipe solutions)
- those that prevent waste production and pollution

CH21-27

Notes

Notes

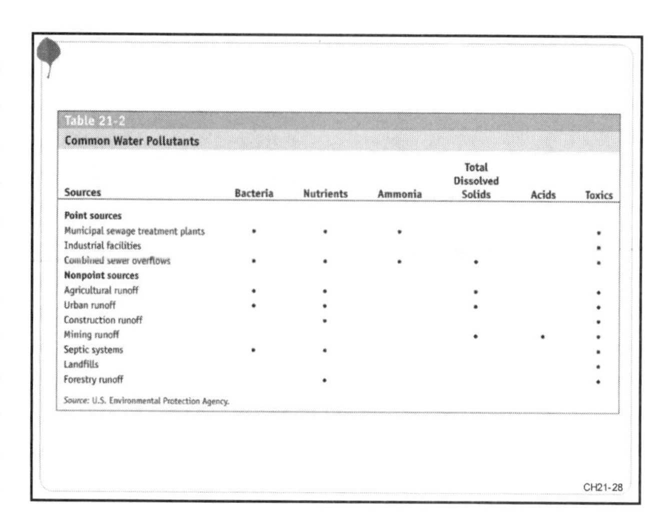

Table 21-2

Common Water Pollutants

Sources	Bacteria	Nutrients	Ammonia	Total Dissolved Solids	Acids	Toxics
Point sources						
Municipal sewage treatment plants	•	•	•			•
Industrial facilities						•
Combined sewer overflows	•	•	•	•		•
Nonpoint sources						
Agricultural runoff	•	•		•		•
Urban runoff	•	•		•		•
Construction runoff		•				•
Mining runoff				•	•	•
Septic systems	•	•				•
Landfills						•
Forestry runoff		•				•

Source: U.S. Environmental Protection Agency.

CH21-28

❖ Legislative Controls

- Legislation to address water pollution has focused on point sources – primarily factories and sewage treatment plants.
- Gains made in controlling such sources have often been offset by increasing levels of pollution from nonpoint sources such as:
  - city streets
  - lawns
  - farm fields

CH21-29

❖ Controlling Nonpoint Pollution

- In the United States, efforts to control nonpoint water pollution are still in their infancy.
- They are gaining popularity because they are often economical solutions that offer other benefits as well.
- The U.S. has focused more on groundwater pollution than nonpoint water pollution because groundwater is an important source of drinking water.

CH21-30

❖ Water Pollution Control Technologies: End-of-Pipe Approaches

- Sewage treatment plants receive waste from many sources, including:
  - Homes
  - Office buildings
  - Factories
  - Hospitals
  - Stormwater drainage systems

CH21-31

❖ Water Pollution Control Technologies: End-of-Pipe Approaches

- Primary treatment is a means of filtering out large objects and solid organic matter.
- Secondary treatment further destroys organic matter and removes much of the nitrogen and phosphorus in sewage using bacteria and other decomposers.
- Tertiary treatment brings sewage wastewater near to drinking water quality but is rarely used because it is so costly.

CH21-32

Table 21-4
Removal of Pollutants by Sewage Treatment Plants

Substance	Percentage Removed by Treatment: Primary	Primary and Secondary
Solids	60	90
Organic wastes	30	90
Phosphorus	0	30
Nitrates	0	50
Salts	0	5
Radioisotopes	0	50
Pesticides	0	0

CH21-33

Notes

Notes

❖ Sustainable Solutions for Water Pollution

- Measures that will collectively serve to reduce our production of water pollutants include:
  - reducing consumption
  - recycling materials
  - reducing industrial waste and municipal sewage
  - using renewable resources
  - stabilizing population growth

CH21-34

Notes

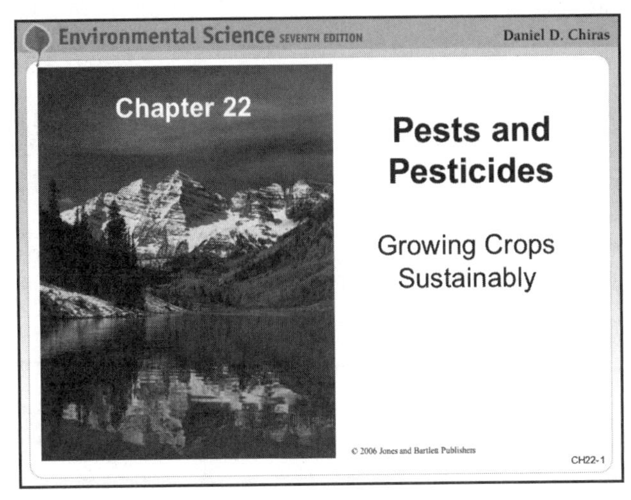

22.1 Chemical Pesticides

- Pest control measures have been used for centuries.
- Earliest techniques included:
  - cultural controls, such as burning fields to kill locusts
  - chemical controls, the use of highly toxic chemicals to kill insect pests
- Early chemicals have been abandoned because they were toxic, ineffective, or environmentally persistent.

CH22-2

❖ Modern Chemical Pesticides

- The first synthetic organic chemical used as a pesticide was a chemical called DDT.
- The development of DDT ushered in a new era of synthetic chemical pesticides.
- Chemical pesticides are either:
  - broad-spectrum substances that kill a variety of pest species
  - narrow-spectrum chemicals that are effective against one or a few pests

CH22-3

Notes

❖ Modern Chemical Pesticides (cont.)

- Three main types of synthetic organic chemical pesticides have been developed over the years:
  - Chlorinated hydrocarbons
  - Organophosphates
  - Carbamates
- These are all neurotoxins.
- Carbamates are less persistent in the environment or body fat of organisms, thus are less likely to be biomagnified in the food chain than organochlorines.
- Unfortunately, they are much more toxic to people.

CH22-4

❖ Growth in the Use of Chemical Pesticides

- The more developed nations, including the U.S., are the major consumers of chemical pesticides.
- In the U.S., the bulk of pesticides are herbicides.

Millions of pounds of active ingredient
1200
1000
800
600
400
200
0
Herbicides
Insecticides
Fungicides
Other Conventional
Other
1980
1990
2000
Year

Fig. 22-2 A profile of pesticide use

CH22-5

❖ Overuse

- Pesticides are often applied far in excess of what is needed by farmers and especially homeowners.
- Homeowners often fail to take precautions to protect themselves from exposure.

CH22-6

Notes

❖ Biological Impacts of Pesticides

- The initial success of DDT led to a rapid increase in pesticide use and in the number of new chemical pesticides.
- One impact of pesticides is that they often kill natural pest control agents such as predatory insects that are more sensitive to pesticides.
- This causes the growth of pest populations.
- Pesticides also kill other beneficial insects such as pollinators.

CH22-7

❖ Biological Impacts of Pesticides (cont.)

- Pesticide use results in the formation of genetically resistant pest species.
- Higher doses and new pesticides used to combat them only worsen the situation by creating even more resistant species.

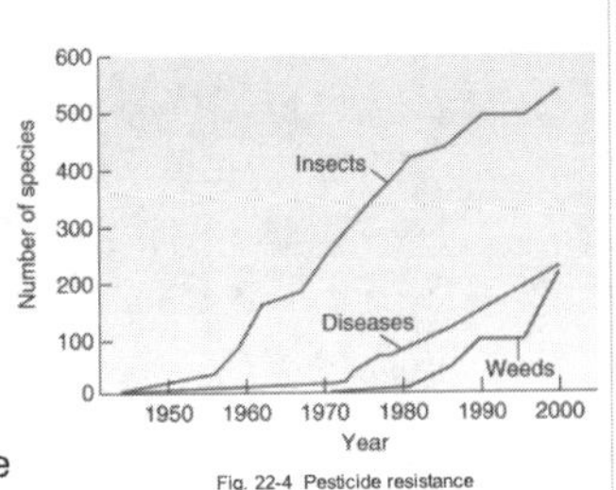

Fig. 22-4 Pesticide resistance

CH22-8

❖ Biological Impacts of Pesticides (cont.)

- Pesticides poison fish, birds, and other species.
- They biomagnify in the food chain.
- Low concentrations in the environment can result in high levels in organisms in the uppermost trophic level.

CH22-9

Notes

❖ Biological Impacts of Pesticides (cont.)

- Pesticides contaminate many foods.
- They have been found in human body tissues even in remote areas of the world, indicating that pesticides are globally distributed.
- Chemical and farm workers are frequently exposed to the highest levels.
- The effects of pesticide exposure range from mild neurological problems to death, depending on the exposure level and type of chemical.

CH22-10

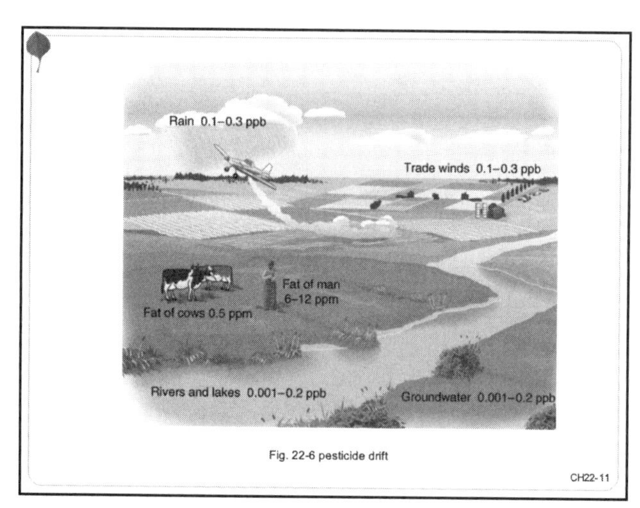

Fig. 22-6 pesticide drift

CH22-11

❖ The Economic Costs of Pesticide Use

- Pesticide use is not just a threat to human health and the environment, it causes considerable economic damage.

CH22-12

Notes

❖ Herbicides in Peace and War

- Auxins are hormones that stimulate plant growth.
- Two widely studied herbicides are auxin-like compounds.
- Herbicides provide many benefits but also have social and environmental costs.
- Use of chemical defoliants during the Vietnam War impacted the health of soldiers and their offspring.

❖ The Alar Controversy: Apples, Alar, and Alarmists?

- Children may be at higher risk to pesticides than adults, especially pesticides used on fruit.

CH22-13

22.2 Controlling Pesticide Use

❖ Bans on Pesticide Production and Use

- Many environmentally harmful pesticides have been banned in the U.S. in the past 30 years.
- Some of these pesticides, however, continue to be used in other countries where they:
  - poison wildlife and migratory birds
  - contaminate crops destined for market

CH22-14

❖ Registering Pesticides

- The EPA registers newly developed and previously introduced pesticides for general or restricted use.
- The registration process stipulates what crops the pesticides can be used on.
- Many improvements are needed in this system to make it more effective, especially more rigorous education and testing of users.

CH22-15

Notes

❖ Establishing Tolerance Levels and Monitoring Produce

- The EPA also sets tolerance levels – acceptable levels of pesticides on food.
- Tolerance levels are monitored by the U.S. Food and Drug Administration.
- Inadequate funding prohibits thorough sampling.

CH22-16

22.3 Integrated Pest Management: Protecting Crops Sustainably

- Integrated pest management is a set of alternative strategies to control pest populations at manageable levels.
- It offers substantial social, economic, and environmental benefits and is the cornerstone of a sustainable system of agriculture.

CH22-17

❖ Environmental Controls

- Environmental controls seek to change both the biotic and abiotic conditions of crops.
- These changes reduce the growth of pest populations, while causing little if any damage to the crop itself.

Fig. 22-9 Strip cropping

CH22-18

❖ Environmental Controls (cont.)

- Planting several different crops on a farm and rotating crops in fields are very effective means of reducing the growth of insect populations without pesticides.
- Crops can be planted after the emergence of insect pests so that the pests have little to eat and subsequently perish.
- Insects are sensitive to nutrient levels in plants, so raising or lowering soil nutrients can effectively control some species.

CH22-19

❖ Environmental Controls (cont.)

- Adjacent weeds or crops may harbor beneficial or harmful insects.
- Controlling what grows beside a given crop can therefore be a powerful tool in protecting crops, with little or no pesticide use.
- Many natural biological control agents can be used to prevent the outbreak of pest species.

Fig. 22-10 Controlling a prickly pear cactus invasion with cactus-feeding moths.

CH22-20

❖ Genetic Controls

- Many insects breed only once a year.
- The introduction of artificially sterilized males into a pest-infested region will result in an abundance of sterile eggs.

CH22-21

Notes

Notes

❖ Genetic Controls (cont.)

- Plants and animals resistant to pests and diseases can be developed through genetic engineering and artificial selection.
- Bacteria that live on plants can also be engineered to be lethal to insect pests.

CH22-22

❖ Chemical Controls

- Conventional pesticides will remain in use, but they must be screened carefully.
- Those that are used must be administered with caution, in amounts and at a time when they are most effective, to minimize their environmental impact.
- This requires better monitoring of crops and infestations.

Fig. 22-11Wick applicator

CH22-23

❖ Chemical Controls (cont.)

- Scientists and farmers are exploring naturally occurring chemicals to control insect pests.
- This could become the cornerstone of pest management in a sustainable system of agriculture.
- Pheromones are natural sex attractants released by insects.
- They can be used to draw insects into traps or sprayed on fields to confuse males so they cannot find females.

CH22-24

❖ Chemical Controls (cont.)

- Two hormones control the life cycles of insects.
- They can be sprayed on crops to alter these cycles and kill pests.
- To be effective, application must be precisely timed.
- These hormones take time to kill off a pest.
- Plants have evolved many natural insect repellants that can be commercially produced and sprayed on crops.

CH22-25

❖ Cultural Controls

- Insect pests can be controlled by many techniques that do not require the use of chemicals.
- These are called cultural controls. They include:
  - noisemakers to frighten birds from crops
  - manual removal of insects
  - quarantines on imported food

CH22-26

❖ Educating the World About Alternative Strategies

- Educating farmers and others about alternative strategies is imperative if sustainable pest management is to become widely adopted.
- Several avenues are available for this important task:
  - universities
  - agricultural agencies
  - farmers' groups
  - nonprofit organizations

CH22-27

Notes

Notes

❖ Government Actions to Encourage Sustainable Agriculture

❖ Governments can promote sustainable agriculture by
- providing low-cost insurance
- develop organic certification programs

CH22-28

Chapter 23: Hazardous and Solid Wastes: Sustainable Solutions

Notes

Environmental Science SEVENTH EDITION Daniel D. Chiras

Chapter 23

Hazardous and Solid Wastes

Sustainable Solutions

CH23-1

23.1 Hazardous Wastes: Coming to Terms with the Problem

- Hazardous wastes come from a variety of sources, among them factories and our homes.
- These toxic materials are signs of unsustainable practices.

Fig. 23-1 Toxic hot spot in Stratford, CT

CH23-2

- Love Canal: The Awakening
  - Love Canal was a hazardous-waste disposal site in Niagara Falls, New York.
  - Leakage from the site caused serious health problems in residents living near it.
  - The incident alerted the public and government officials to the problem of improper hazardous-waste disposal.

CH23-3

Notes

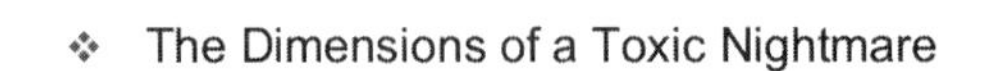

❖ The Dimensions of a Toxic Nightmare

- The problem with hazardous wastes is twofold.
  1. Tens of thousands of hazardous-waste sites are in need of cleanup around the world.
     - These sites contaminate groundwater and can affect the health of people who live nearby.
  2. Factories continue to generate millions of tons of hazardous waste each year.

CH23-4

❖ LUST—It's Not What You Think

- **L**eaking **U**nderground **S**torage **T**anks
- Hundreds of thousands of underground storage tanks have been installed in industrial nations.
- They are used to store many potentially toxic substances.
- Over time, steel tanks corrode and begin to leak, contaminating groundwater used for cooking, drinking, and bathing.

CH23-5

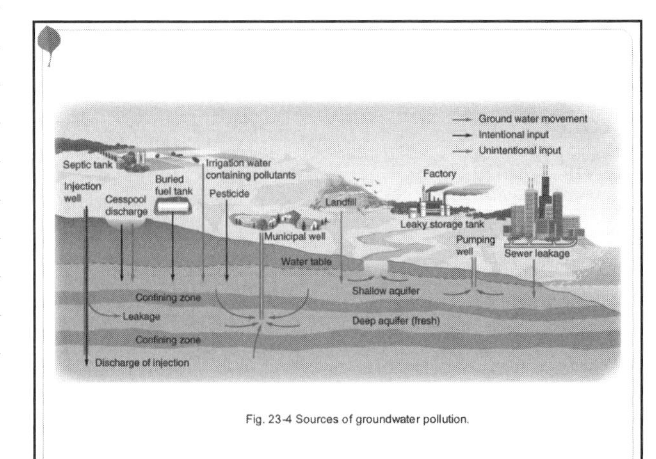

Fig. 23-4 Sources of groundwater pollution.

CH23-6

Notes

23.2 Managing Hazardous Wastes

- Addressing the issue of hazardous waste requires:
  - plans to clean up contaminated sites
  - preventive actions to greatly reduce or eliminate hazardous-waste production

CH23-7

❖ The Superfund Act: Cleaning Up Past Mistakes

- The Superfund Act provides money to clean up hazardous waste dumps and other contaminated sites.
- This money comes from a tax on oil and petrochemicals and is replenished by fees charged to those responsible for the contamination.

CH23-8

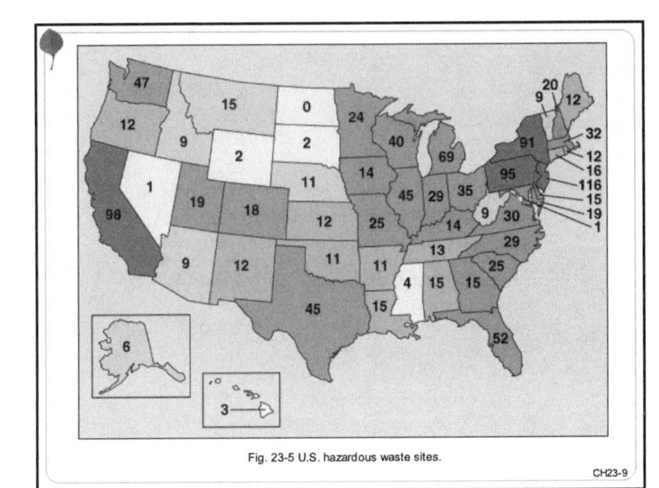

Fig. 23-5 U.S. hazardous waste sites.

CH23-9

Notes

❖ The Superfund Act: Cleaning Up Past Mistakes (cont.)

- The cleanup of hazardous waste sites under the Superfund Act has proven extremely slow, costly, and litigious.
- The Superfund Act has been criticized because it provides money for cleaning up property but no compensation for health damage.
- Many of the cleanups are considered inadequate.

CH23-10

❖ The Superfund Act: Cleaning Up Past Mistakes (cont.)

- Rather than spending millions of dollars to identify responsible parties, the Superfund program might work better with a no-fault policy.
- Such a policy would provide funds to clean up sites regardless of who is liable.

CH23-11

❖ What to Do with Today's Waste: Preventing Future Disasters

- The Resource Conservation and Recovery Act established a reporting system to monitor hazardous wastes from production through disposal.
- It seeks to eliminate illegal and improper hazardous waste disposal.

CH23-12

Notes

❖ What to Do with Today's Waste: Preventing Future Disasters (cont.)

- Additional steps are needed to further reduce environmental contamination by hazardous wastes.
- The definition of what is hazardous must be broadened to include:
  - municipal waste
  - sewage
  - pesticides
  - mine waste

CH23-13

❖ What to Do with Today's Waste: Preventing Future Disasters (cont.)

- Tighter regulations for the disposal of hazardous wastes have led to commendable efforts to reduce waste production by some companies.
- They have also led to illegal dumping by less scrupulous companies.
- Some companies export hazardous wastes to LDCs with little or no oversight of such practices.

CH23-14

❖ Dealing with Today's Wastes: A Variety of Options

- Many options exist for getting rid of waste.
- The most sustainable approaches involve steps that reduce or eliminate hazardous-waste output.
- You don't have waste if you don't make it.

CH23-15

Notes

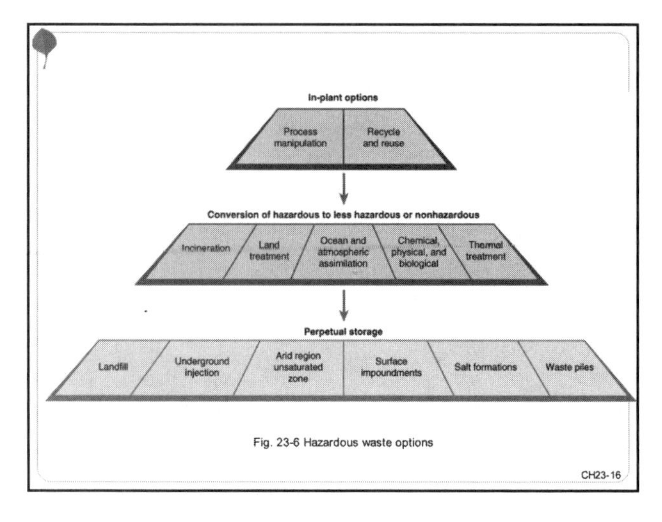

Fig. 23-6 Hazardous waste options

CH23-16

❖ Dealing with Today's Wastes: A Variety of Options (cont.)

- Changes in manufacturing processes such as substitution and process manipulation are often the simplest and most cost-effective means of reducing or eliminating hazardous wastes.
- Waste output can also be dramatically reduced by recycling and reusing wastes.

CH23-17

❖ Dealing with Today's Wastes: A Variety of Options (cont.)

- Hazardous wastes can be converted to nontoxic or less toxic substances by chemical, physical, and biological means, such as:
  - neutralization
  - combustion
  - low-temperature thermal decomposition
  - bacterial decay

CH23-18

❖ Dealing with Today's Wastes: A Variety of Options (cont.)

- Not all waste can be eliminated by prevention, recycling, or detoxification.
- Perpetual storage remains the final, yet least sustainable, option.

CH23-19

❖ Disposing of Radioactive Wastes

- Disposing of radioactive waste will be a problem for a long time, even though nuclear power and nuclear weapons production are on the decline.
- Safe, permanent repositories are needed to store huge amounts of waste produced by power plants and other facilities.

Fig. 23-8 Aerial view of Yucca Mountain, Nevada

CH23-20

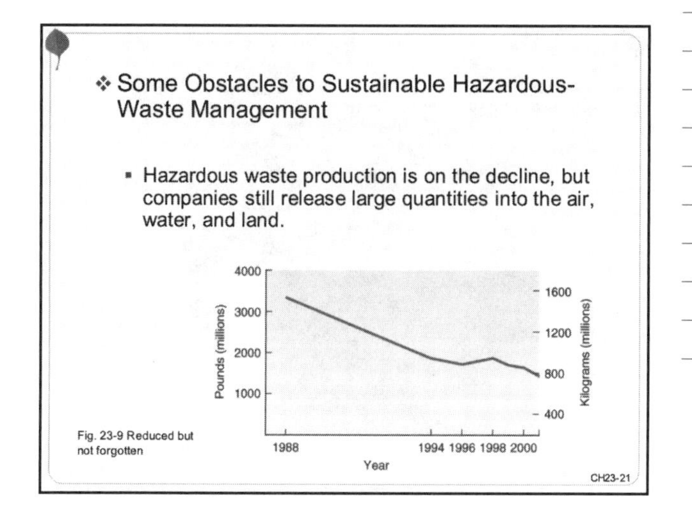

Fig. 23-9 Reduced but not forgotten

Notes

Notes

❖ Individual Actions Count

- Individuals can help reduce the hazardous-waste threat by:
  - properly disposing of hazardous materials
  - avoiding their use when possible
  - cutting down on nonessential consumption

CH23-22

23.3 Solid Wastes: Understanding the Problem

- More developed nations produce enormous amounts of solid waste each year.
- Much of the waste is burned or landfilled.
- Waste production is increasing in many countries such as the U.S.
- Recovery rates (recycling and composting) are growing much faster, a trend that bodes well for the future.

CH23-23

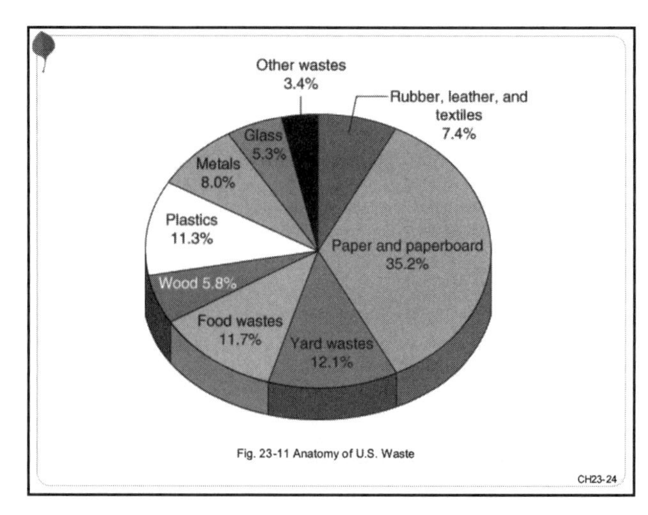

Fig. 23-11 Anatomy of U.S. Waste

CH23-24

23.4 Solving a Growing Problem Sustainably

- Solid waste management strategies fall into one of three categories:
  1. those that deal with waste after it has been produced
  2. those that divert waste back into the production-consumption cycle
  3. those that prevent waste generation

Fig. 23-12 Strategies for reducing waste.

CH23-25

❖ The Traditional Strategy: The Output Approach

- Worldwide, most trash is still dumped in landfills.
- Open garbage dumps have been replaced by sanitary landfills.
- Landfills are pits into which garbage is dumped, then covered daily with a thin layer of dirt to eliminate odors and pests.

CH23-26

❖ The Traditional Strategy: The Output Approach (cont.)

- Oceans have long been viewed as a huge waste dump site for a variety of wastes, including radioactive materials.
- This practice is no longer legal in the U.S., but it continues in other countries.

Fig. 23-13 Ocean dumping of sewage has declined.

CH23-27

Notes

Notes

- The Traditional Strategy: The Output Approach (cont.)
  - Garbage can be burned in incinerators, which greatly reduces trash.
  - This, however, even when linked to energy production, is viewed by many as an unsustainable way of dealing with municipal solid waste.

CH23-28

- Sustainable Options: The Input Approach
  - Source reduction techniques reduce the amount of waste entering the waste stream.
  - They represent the most sustainable waste management strategy.
  - More durable goods last longer and reduce the amount of waste.
  - Manufacturers and consumers can play large roles by making and buying more durable products.

CH23-29

- Sustainable Options: The Input Approach (cont.)
  - Efforts to make products smaller and more compact can significantly reduce resource demand.
  - One of the most effective means of reducing solid waste is to reduce consumption – buy only what you need.

CH23-30

❖ The Throughput Approach: Reuse, Recycling, and Composting

- The throughput approach diverts waste from the waste stream for recycling and reuse.
- Reusing materials and products cuts down on resource demand.
- Citizens can participate in two ways:
  - donating used products
  - buying used goods

CH23-31

Table 23-1

Reuse and Recycling of Solid Wastes

Material	Reuse and Recycling
Paper	Repulped and made into cardboard, paper, and a number of paper products. Incinerated to generate heat. Shredded and used as mulch or insulation.
Organic matter	Composted and added to gardens and farms to enrich the soil. Incinerated to generate heat.
Clothing and textiles	Shredded and reused for new fiber products or burned to generate energy. Donated to charities or sold at garage sales.
Glass	Returned and refilled. Crushed and used to make new glass. Crushed and mixed with asphalt. Crushed and added to bricks and cinder blocks.
Metals	Remelted and used to manufacture new metal for containers, buildings, and other uses.

Source: Modified from B. J. Nebel (1981). *Environmental Science*. Englewood Cliffs, NJ: Prentice-Hall, p. 297.

CH23-32

❖ The Throughput Approach (cont.)

- Many products can be returned to recycling facilities, where they are shipped to factories to be used to make new products.
- Recycling is on the rise, but most countries have barely tapped the full potential of recycling.
- Many states now recycle 30% or more of their municipal solid waste.

CH23-33

Notes

Notes

❖ The Throughput Approach (cont.)

- Recycling programs generally involve drop-off sites or curbside pickup.
- These may be run by private industry or by city and town governments.
- Although recycling is a promising strategy for waste management, many obstacles hinder its full implementation.
- Some of the these are federal laws and subsidies that give the raw materials industry an unfair economic advantage over recycling.

CH23-34

❖ The Throughput Approach (cont.)

- Expanding the amount of garbage that is recycled requires removal of:
  - legal and economic barriers
  - subsidies that give benefits to companies that use virgin materials rather than recycled ones.
- Commitment on the part of citizens is vital to making recycling a success.

CH23-35

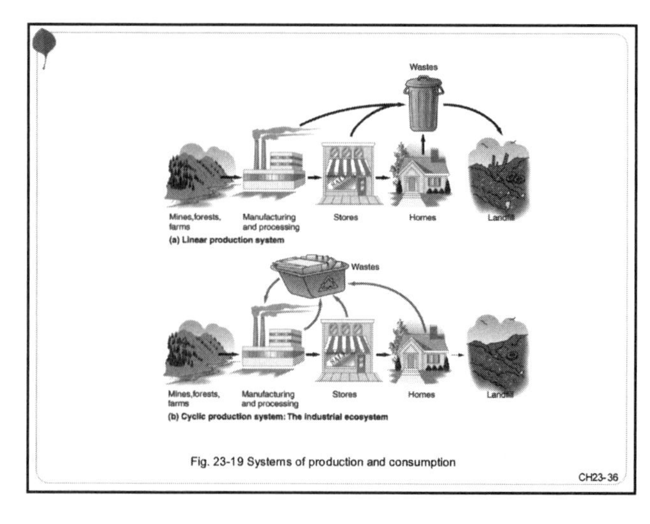

Fig. 23-19 Systems of production and consumption

CH23-36

❖ The Throughput Approach (cont.)

- One of the most important boosts to recycling is purchasing products made from recycled materials.
- Governments, businesses, and individuals can all engage in this activity.
- Governments can help strengthen the markets by:
  - providing incentives to companies that use recycled materials
  - requiring them to use a certain percentage of recycled content in their products

CH23-37

❖ The Throughput Approach (cont.)

- Composting is a way of returning the nutrients contained in organic matter such as yard waste and food scraps to the soil.
- This strategy not only reduces the need for landfilling, it helps nourish soils, creating a more closed-loop system.
- Recycling is not only good for the environment, it often costs less than other strategies and creates many jobs.

CH23-38

Notes

Chapter 24: Environmental Ethics: The Foundation of a Sustainable Society

Notes

24.1 The Frontier Mentality Revisited

- The frontier ethic has long historical roots and is deeply embedded in modern civilization.
- It is a notion of:
  - unlimited resources for exclusive human use
  - humans being apart from nature, succeeding by dominating it
- It influences profoundly the way we meet our needs, and how we view and solve environmental problems of great urgency.

CH24-2

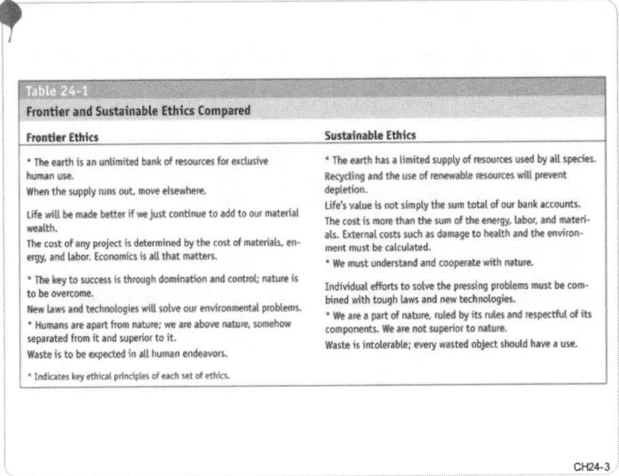

Table 24-1
Frontier and Sustainable Ethics Compared

Frontier Ethics	Sustainable Ethics
* The earth is an unlimited bank of resources for exclusive human use.	* The earth has a limited supply of resources used by all species.
When the supply runs out, move elsewhere.	Recycling and the use of renewable resources will prevent depletion.
Life will be made better if we just continue to add to our material wealth.	Life's value is not simply the sum total of our bank accounts.
The cost of any project is determined by the cost of materials, energy, and labor. Economics is all that matters.	The cost is more than the sum of the energy, labor, and materials. External costs such as damage to health and the environment must be calculated.
* The key to success is through domination and control; nature is to be overcome.	* We must understand and cooperate with nature.
New laws and technologies will solve our environmental problems.	Individual efforts to solve the pressing problems must be combined with tough laws and new technologies.
* Humans are apart from nature; we are above nature, somehow separated from it and superior to it.	* We are a part of nature, ruled by its rules and respectful of its components. We are not superior to nature.
Waste is to be expected in all human endeavors.	Waste is intolerable; every wasted object should have a use.

* Indicates key ethical principles of each set of ethics.

CH24-3

Notes

24.2 Sustainable Ethics: Making the Transition

- The frontier ethic may be replaced by a more sustainable view as evidence of global decline and its wide-ranging effects continues to mount.

CH24-4

❖ Leopold's Land Ethic: Planting the Seed

- Aldo Leopold described a land ethic that called on people to:
  - view themselves as a part of the environment
  - discard the notion of humans as conquerors of nature

CH24-5

❖ A New View to Meet Today's Challenges: Sustainable Ethics

- Sustainable ethics sees the world as a finite supply of resources shared by all organisms.
- It views people as a part of nature.
- It maintains that they will succeed best by fitting human systems within the limits and capacities of natural systems.
- This ethical system may inspire restraint and ecological design.

CH24-6

- Toward a Humane, Sustainable Future
  - Worldwide adoption of sustainable ethics is necessary to change the way we view the Earth and one another.

CH24-7

24.3 Developing and Implementing Sustainable Ethics

- Many avenues for changing outmoded ethical systems are available to us.
- Promoting Models of Sustainability
  - Publicizing models of sustainable action offers inspiration and practical examples of what individuals, businesses, and governments can do to build a sustainable future.

CH24-8

- Education
  - All levels of education can be enlisted in the effort to foster an understanding of ethics and actions that will contribute to a sustainable future.

CH24-9

Notes

Notes

❖ Religious Organizations

- These are an established forum for education whose efforts could be partially dedicated to the exploration of sustainable ethics and lifestyles.

CH24-10

❖ Declarations of Sustainable Ethics and Policy

- Official declarations by states, nations, and groups of nations can help promote widespread adoption of sustainable ethics and actions.

CH24-11

❖ A World Organization Dedicated to Sustainable Development

- The United Nations Commission on Sustainable Development is an international organization dedicated to promoting sustainable development.

CH24-12

❖ A Role for Everyone

- Participation by all sectors of society is essential to building a sustainable society.

CH24-13

24.4 Overcoming Obstacles to Sustainability

- Many obstacles lie on the road to sustainability.
- The process will likely take a century or more.

CH24-14

❖ Faith in Technological Fixes

- The belief that technology can solve all of the world's problems hinders:
  - development of a sustainable ethic
  - implementation of simple, cost-effective sustainable practices

CH24-15

Notes

Notes

❖ Apathy, Powerlessness, and Despair

- Apathy and feelings of powerlessness and despair hinder progress toward sustainability.
- Many examples show that individuals can be truly powerful in changing the world.

Fig. 24-5 Getting active can help reduce one's feelings of despair.

CH24-16

❖ The Self-Centered View

- Efforts to foster individual action and responsibility are hindered by:
  - concern for the self
  - feelings of insignificance
  - cash conscience
  - blame shifting

CH24-17

❖ Ego Gratification

- Feelings of inadequacy are often offset by the accumulation of material possessions.
- Over-consumption is a key factor in the decline of the environment.

CH24-18

❖ Economic Self-Interest and Outmoded Governmental Policies

- Unsustainable economic systems and governments that support them are major barriers to sustainable development.

CH24-19

24.5 Sustainable Ethics: How Useful Are They?

- Values affect the way we act.
- To some people, the value of things is based on their usefulness.
- Sustainable ethics may be one of the most useful sets of values because it serves people, the environment, and the economy much better than shortsighted, outmoded systems of belief.

CH24-20

Notes

Chapter 25: Sustainable Economics: Understanding the Economy and Challenges Facing the Industrial Nations

Notes

Environmental Science SEVENTH EDITION — Daniel D. Chiras

Chapter 25

Sustainable Economics

Understanding the Economy and Challenges Facing the Industrial Nations

CH25-1

25.1 Economics, Environment, and Sustainability

- The human economy has been functioning for thousands of years.
- Recently people have begun to think seriously about developing a sustainable economy.
- It should serve people equitably and protect and enhance the environment upon which we all depend.

CH25-2

❖ Economic Systems

- Two types of economic systems exist:
  - command economies
  - market economies
- Command economies are run in large part by governments.
- In market economies, decisions about production are largely determined by prices and by people's ability to pay.
- Both types of economies have had enormous environmental impact.

CH25-3

Notes

❖ The Law of Supply and Demand

- The law of supply and demand shows that prices affect both:
  - supply of goods and services
  - demand for goods and services
- The prices of goods and services in a market economy are largely determined by the interaction of supply and demand.

CH25-4

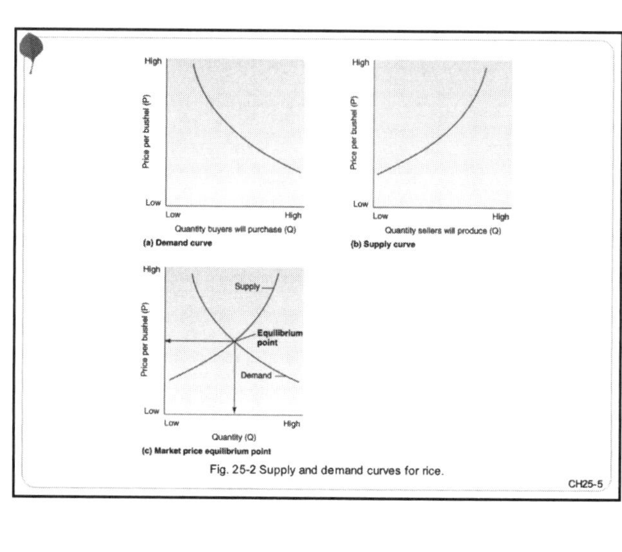

Fig. 25-2 Supply and demand curves for rice.

CH25-5

❖ Environmental Implications of Supply and Demand

- Economic considerations have profound impacts on activities that affect the quality of our environment and the sustainability of our society.

CH25-6

Notes

❖ Measuring Economic Success: The GNP

- The gross national product is a measure of the economic output of a nation, including all goods and services.
- It is used to track the success of economies.

CH25-7

25.2 The Economics of Pollution Control

- Pollution creates outside costs called externalities.
- Externalities are borne by society at large, not by producers.
- Pollution control technologies reduce the emission of harmful substances and thus reduce external costs.
- They also add to the cost of producing goods and services.

CH25-8

❖ Cost-Benefit Analysis and Pollution Control

- Cost-benefit analysis answers the question:

"How much should be spent to reduce pollution to a level at which the costs of control equal the benefits?"

(reduced externalities)

- Sustainable approaches can reduce pollution emissions at a far lower cost than end-of-pipe solutions.

CH25-9

Notes

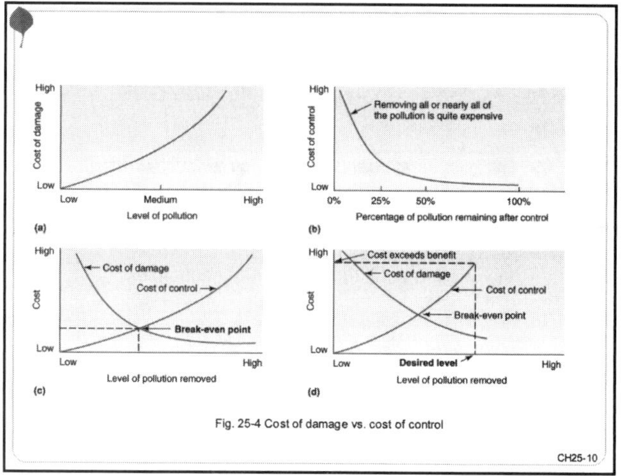

Fig. 25-4 Cost of damage vs. cost of control

CH25-10

❖ Who Should Pay for Pollution Control?

- Pollution controls and other environmental protection strategies are paid for by:
  - consumers if the costs are borne by businesses
  - taxpayers if the costs are shouldered by government

CH25-11

❖ Does Pollution Control Always Cost Money?

- Reducing or eliminating pollution can be a profitable venture that adds to the bottom line of companies.
- Pollution prevention and other techniques often save companies considerable sums of money, especially if all costs are calculated.

Fig. 25-5 Alaskan oil spill

CH25-12

25.3 The Economics of Resource Management

- The management of natural resources is affected by numerous economic factors.

CH25-13

- Time Preference
  - Resource management is influenced by time preference, one's temporal preference for earnings.
  - Time preference is affected by current needs, uncertainty, and inflation.

CH25-14

- Opportunity Cost
  - Money can be put to many uses.
  - Many people choose options that provide the highest returns.
  - That may not include investment in wise resource management.

CH25-15

Notes

Notes

❖ Discount Rates

- Discount rates are used to calculate the present value of different options in resource management.
- Decisions based on the discount rate tend to emphasize immediate returns.
- This results in the liquidation of natural resources rather than their sustainable harvest.

CH25-16

❖ Ethics

- Not all decisions about resource management are based on economics.
- Ethics can play a big role in determining actions, overriding other immediate concerns such as opportunity costs.

CH25-17

25.4 What's Wrong with Economics: An Ecological Perspective

- The economic system has several key flaws when viewed from an ecological perspective.
- Correcting these flaws can help us create a sustainable human economy.

CH25-18

❖ Economic Shortsightedness

- Economic systems and the participants in them often fail to take into account long-term supplies.
- This is a dangerous trend that results in an underpricing of many natural resources, leading to environmentally unsustainable activities.
- Several mechanisms are available to incorporate such concerns into economic decision making, including user fees.

CH25-19

❖ Economics and Growth

- Continual economic growth is the abiding principle of many economies and the measure of their success.
- It is ultimately incompatible with the finite world in which we live.
- Economic growth is fueled by population growth and ever-increasing per capita consumption.

CH25-20

❖ Growth and the GNP: Some Fundamental Flaws

- The GNP is our measure of success in continued economic growth.
- This is an inaccurate measure of the welfare of a nation's people.
- It fails to distinguish economic activity that enhances our welfare from that which results in a decreased quality of life.

CH25-21

Notes

Notes

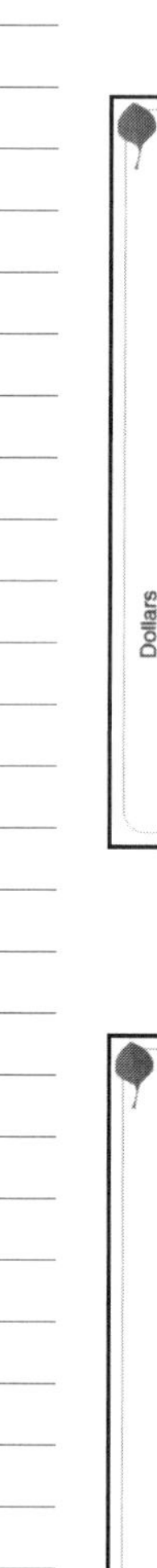

❖ Growth and the GNP: Some Fundamental Flaws (cont.)

- Measures of economic performance that account for economic negatives provide a more accurate picture of the welfare of nations.
- More precise information about the welfare of people comes from tracking key social, economic, and environmental trends.

CH25-22

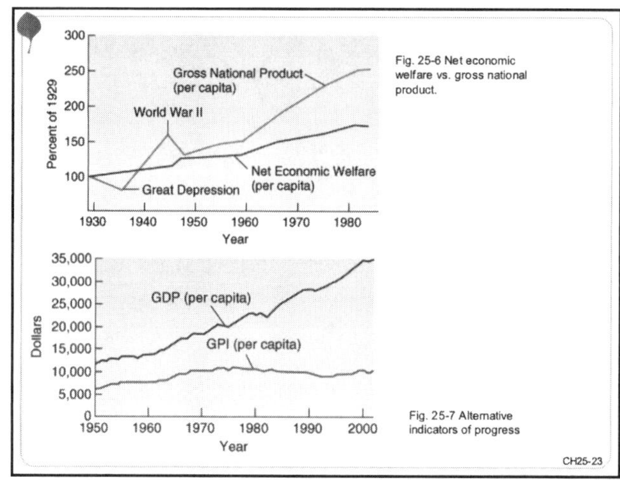

Fig. 25-6 Net economic welfare vs. gross national product.

Fig. 25-7 Alternative indicators of progress

CH25-23

❖ Growth and the GNP: Some Fundamental Flaws (cont.)

- Applying principles of sustainability can help ensure that economic activity translates into improvements in quality of life.
- Creating a sustainable future may require us to maintain and improve human well-being without increasing economic throughput.
- Many sustainable strategies, such as efficiency, permit the attainment of this goal.

CH25-24

❖ Fostering Local and Regional Self-Reliance

- For most of us, trade is viewed as a highly desirable activity.
- The interdependence it creates has many adverse effects on our long-term sustainability.
- Greater local and regional self-reliance, while controversial, may be essential to achieving a sustainable future.

CH25-25

❖ Creating an Ecologically Compatible Society

- The practices of sustainability can create a society that is
  - more efficient
  - less resource intensive
  - more ecologically compatible

CH25-26

25.5 Creating a Sustainable Economic System: Challenges in the Industrial World

❖ Numerous changes can help us forge a sustainable economy.

❖ It must meet human needs without foreclosing on future generations.

- It must not destroy the natural resource base that makes all economic activity possible.

CH25-27

Notes

Notes

Table 25-1

Characteristics of a Sustainable Economy

Participants take a long-term systems view that

- Recognizes the importance of natural systems to human well-being.
- Ensures that economic activities improve the quality of life of all, not just a select few.
- Seeks ways to ensure that economic activities maintain or improve natural systems.
- Uses all resources extremely efficiently.
- Promotes maximum recycling and reuse.
- Relies heavily on clean, renewable technologies.
- Restores damaged ecosystems.
- Promotes regional self-reliance.
- Relies on appropriate technology.
- Designs and builds using principles of ecological sustainability.

CH25-28

- Harnessing Market Forces to Protect the Environment
  - Many market mechanisms can be applied to environmental problems, allowing businesses to:
    - innovate
    - save money
    - reduce the burden of regulations

CH25-29

- Harnessing Market Forces to Protect the Environment (cont.)
  - Green taxes are levies on undesirable products or activities.
  - They create a disincentive to companies, spurring interest in finding environmentally sustainable alternatives.
  - Governments can use grants and tax incentives to encourage sustainable business practices and products.

CH25-30

Notes

- Harnessing Market Forces to Protect the Environment (cont.)
  - Permits that regulate the amount of pollution facilities are allowed to release can be bought and sold.
  - This provides companies with an economic incentive to reduce pollution.
  - If they can find a cost-effective way to reduce emissions, they can sell their unused permitted emissions.

CH25-31

- Harnessing Market Forces to Protect the Environment (cont.)
  - Governments often subsidize or give economic advantage to unsustainable practices.
  - Removing these market barriers can create a level economic playing field.
  - This will permit sustainable activities and products to flourish.

CH25-32

- Corporate Reform: Greening the Corporation
  - Companies can be forced to become sustainable through regulations and market mechanisms.
  - The nature of business is impacted by individual responsibility and action by:
    - business owners
    - members of boards of directors
    - CEOs
    - employees

CH25-33

Notes

❖ Green Products and Green Seals of Approval

- Products vary in the degree to which they contribute to sustainability.
- By purchasing green products, individuals help promote a sustainable economy.
- Product labeling programs can help individuals select the most environmentally sustainable products.

GREEN SEAL

Fig 25-10 The Green Seal of approval

CH25-34

❖ Appropriate Technology and Sustainable Economic Development

- Appropriate technologies are an essential part of a sustainable future.
- They rely on local resources, are efficient, and produce little pollution.

Table 25-2

Characteristics of Appropriate Technology

- Machines are small to medium sized.
- Human labor is favored over automation.
- Machines are easy to understand and repair.
- Production is decentralized.
- Factories use local resources.
- Factories use renewable resources whenever possible.
- Equipment uses energy and materials efficiently.
- Production facilities are relatively free of pollution.
- Production is less capital intensive than conventional technology.
- Management stresses meaningful work, allowing workers to perform a variety of tasks.
- Products are generally for local consumption.
- Products are durable.
- The means of production are compatible with local culture.

CH25-35

❖ A Hopeful Future

- A sustainable economy:
  - conforms to ecological design principles
  - is dynamic
  - is full of opportunity

CH25-36

25.6 Environmental Protection versus Jobs: Problem or Opportunity?

- Environmental protection may be a stimulant to economic progress, not an impediment.
- Many sustainable strategies result in cost-competitive goods and services.
- This can create more jobs with little environmental impact.
- A sustainable economy will involve a major employment shift.

CH25-37

Notes

Chapter 26: Sustainable Economic Development: Challenges Facing the Developing Nations

Notes

Environmental Science SEVENTH EDITION — Daniel D. Chiras

Chapter 26

Sustainable Economic Development

Challenges Facing the Developing Nations

CH26-1

26.1 Conventional Economic Development Strategies and Their Impacts

- Economic and environmental problems abound in the LDCs.
- They are the result of many interacting forces, including:
  - overpopulation
  - political corruption
  - ineptitude
  - war
  - economic exploitation
  - misguided development efforts

CH26-2

- What Is Wrong with Western Development Assistance?
  - Western-style development efforts often result in means of production that:
    - centralize wealth
    - destroy sustainable lifestyles
  - These efforts radically reshape the environment and destroy the soil and other natural resources.
  - They also promote costly dependency on Western economies.

CH26-3

Notes

❖ Who's Financing International Development?

- Most financial resources come from international banks and government development agencies.
- Most of it is uncoordinated and largely unconcerned with achieving a sustainable form of development.

CH26-4

26.2 Sustainable Economic Development Strategies

- Creating a sustainable future for the LDCs will require many steps, including efforts to:
  - slow or halt the growth of the human population
  - improve education and health care
  - ensure the protection of basic human rights
- The infrastructure of each country will need to be rethought and reshaped.

CH26-5

❖ Employing Appropriate Technology

- Appropriate technology is:
  - environmentally compatible
  - easily understood and repaired
  - reliant on local resources, especially people
- It is an essential component of sustainable development in LDCs.
- These technologies are important to production of food and goods.

CH26-6

❖ Creating Environmentally Compatible Systems of Production

- Production systems designed to fit within the local environment are essential.
- They are more affordable and tend to benefit participants more than large-scale Western development projects.

CH26-7

❖ Tapping Local Expertise and Encouraging Participation

- Knowledge of a sustainable means of production often exists within local communities in LDCs.
- It can provide the basis for much new development.
- Such projects must be deemed important; solutions must enhance the local culture to be successful.

CH26-8

❖ Promoting Flexible Strategies

- Large bureaucracies that attempt to manage development projects lack the flexibility needed to respond successfully to problems.
- Creating flexible management structures that are more responsive to emergent needs will help ensure greater success.

CH26-9

Notes

Notes

❖ Improving the Status and Expanding the Role of Women

- Women play an important role in shaping a country's future.
- They are often relegated to an inferior position.
- Creation of a sustainable future can be enhanced by increasing the status of women, and creating greater opportunities in:
  - Education
  - Employment
  - Credit attainment

CH26-10

❖ Preserving Natural Systems and Their Services

- To be sustainable, social and economic development strategies must protect natural systems.
- Natural systems provide many free services such as flood control and water purification.

CH26-11

❖ Improving the Productivity of Existing Lands

- A vital first step in sustainable economic development is restoration and sustainable enhancement of productivity in:
  - previously or currently used land
  - other resources such as farmland and forests
- This reduces the pressure on undeveloped land.

CH26-12

26.3 Overcoming Attitudinal and Economic Barriers

- Several substantial attitudinal and economic barriers obstruct sustainable development.

CH26-13

❖ Attitudinal Barriers

- Environmental protection and efficiency are often seen as luxuries.
- They are truly essential to the survival and prosperity of people in less developed nations.
- Sustainable development strategies must seek to meet current human needs while:
  - promoting efficiency
  - protecting the life-support systems on which people rely

CH26-14

❖ Economic Barriers

- Less developed nations will require huge sums of money to develop sustainably.
- Several measures could make this possible, including:
  - decreases in military expenditures
  - reduced economic exploitation by the Western world
  - debt relief

CH26-15

Notes

Notes

27.1 The Role of Government in Environmental Protection

- Forms of Government
  - Democratic nations are those in which most decisions are made by elected officials.
  - The economies tend to be controlled by market forces.

CH27-2

- Forms of Government (cont.)
  - Communist nations are those designed around:
    - the notion of public ownership of the means of production
    - the distribution of goods and services according to need, not ability to pay

CH27-3

Notes

❖ Government Policies and Sustainability

- Democratic governments regulate activities, such as environmental protection, via three measures:
  1. tax policy
  2. direct financial support
  3. laws and regulations
- The President also wields enormous power through the executive order.

CH27-4

27.2 Political Decision Making: The Players and the Process

❖ Government Officials

- Government officials are charged with the task of creating laws and regulations and enforcing them for the general good.
- In democratic nations, government officials are either elected or appointed.
- Both types are accountable to the general public.
- Elected officials are generally more responsive, as their reelection depends on satisfying the electorate.

CH27-5

❖ The Public

- Public policy is forged by elected officials with input from the general public.
- Although many people are discouraged with the process and are apathetic, letters and other forms of communication to elected officials can help shape public policy.

Fig. 27-CO

CH27-6

Notes

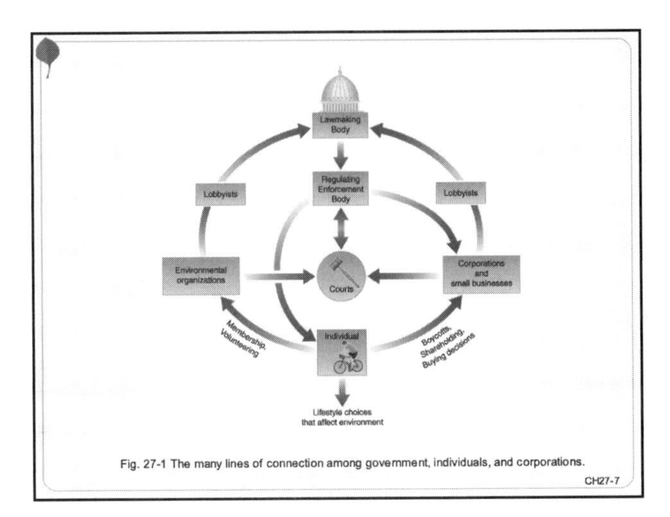

Fig. 27-1 The many lines of connection among government, individuals, and corporations.

CH27-7

❖ Special Interest Groups

- Special interest groups can wield an inordinate amount of political power through:
  - lobbyists
  - financial support of political campaigns
- Nonetheless, citizen interests have, on many occasions, overcome powerful PAC interests.

❖ Environmental Groups

- Environmental organizations take many forms and work in many ways to protect the environment.

CH27-8

27.3 Environment and Law: Creating a Sustainable Future

❖ Evolution of U.S. Environmental Law

- U.S. environmental law began locally.
- It evolved to higher levels of government because local problems spread to other areas and required attention from a higher political authority.

CH27-9

Notes

❖ The National Environmental Policy Act

- The National Environmental Policy Act requires examination for environmental impact of:
  - all projects on federally owned land
  - all federally subsidized projects
- It established the environmental impact statement as a means of assessing potential effects and suggesting ameliorative actions.

CH27-10

❖ The Environmental Protection Agency

- The U.S. Environmental Protection Agency manages and enforces most of the country's environmental laws.
- It sets standards for pollution.
- It conducts research to expand our understanding of the effects of toxic substances and other pollutants on human health and the environment.

CH27-11

❖ Principles of Environmental Law

- Environmental law is based on:
  - statutes and local ordinances (specific laws passed by Congress and other governing bodies)
  - common law (principles derived from thousands of years of legal decisions)

CH27-12

❖ Principles of Environmental Law (cont.)

- Statutory law generally describes broad goals.
- The regulations and enforcement that help us reach these goals come from the EPA and other federal agencies, which have the expertise to draft them.
- Common law is based on time-honored principles and attempts to balance competing societal interests.

CH27-13

❖ Principles of Environmental Law (cont.)

- People can use their property or land in any way they see fit, **if** it does not cause injury or annoyance to others (a nuisance).
- Many environmental lawsuits are based on the nuisance principle.
- Negligence is an unreasonable act that causes personal or property damage.

CH27-14

❖ Principles of Environmental Law (cont.)

- The burden of proof needed to win environmental lawsuits is not always easy to achieve, due to factors such as:
  - inconclusive data connecting pollution with disease
  - long lag times between exposure and effect

CH27-15

Notes

Notes

❖ Resolving Environmental Disputes out of Court

- Out-of-court settlements through mediation provide a less hostile approach to conflict resolution.
- They can reduce or avoid costs of environmental lawsuits for both defendant and plaintiff.

CH27-16

27.4 Creating Governments That Foster Sustainability

- Most governments promote policies and economies that are:
  - shortsighted
  - growth-oriented
  - exploitive
- Correcting perceptions and government policies could help us reshape our unsustainable economy.

CH27-17

❖ Creating Government with Vision

- A long-term perspective is essential to create a sustainable government.
- Research and education are essential to raise the level of awareness of environmental problems and sustainable solutions among public and elected officials.

CH27-18

❖ Creating Government with Vision (cont.)

- In government, long-range planning and action are often sacrificed in the interest of addressing immediate problems.
- This is called **crisis politics**.
- Crisis management tends to apply marginal solutions.
- It allows little problems to become larger and more difficult to solve.

CH27-19

❖ Creating Government with Vision (cont.)

- Longer terms of office, term limitations and shorter campaigns for elected officials could:
  - reduce political shortsightedness
  - encourage solutions that contribute to sustainability

CH27-20

❖ Creating Government with Vision (cont.)

- Creating a sustainable future requires short-term solutions to address immediate problems.
- But, it ultimately hinges on a long-term, proactive approach that seeks to prevent problems.

CH27-21

Notes

Notes

❖ Ending Our Obsession with Growth

- Many measures could help temper the obsession with growth found in many nations, which is so detrimental to the goal of sustainability.

CH27-22

❖ Reducing Exploitation and Promoting Self-Reliance

- Measures to promote other goals of sustainability could help:
  - reduce society's exploitation of natural systems
  - reduce society's exploitation of other people
  - promote greater self-reliance

CH27-23

❖ Reducing Exploitation and Promoting Self-Reliance (cont.)

- Offices of sustainable economic development in state and national governments could replace existing offices of business development.
- They could promote sustainable business development and help current businesses become more sustainable.
- These actions are beneficial to both the Earth and the economy.

CH27-24

Notes

Table 27-1

Summary of Recommendations for Sustainable Government

- Adopt a national sustainable development strategy that promotes conservation, recycling, renewable resource use, restoration, and population control.
- Repeal or modify existing laws that hinder sustainability.
- Increase support of research on environmental problems and sustainable solutions to them.
- Increase support of education on environmental issues and sustainable solutions.
- Pass a national Sustainable Futures Act.
- Establish offices of critical trends analysis at state and national levels.
- Establish offices of sustainable economic development in state governments.
- Evaluate all new and existing policies on the basis of sustainability.
- Adopt sustainable land-use planning at state and national levels.
- Implement policies and actions outlined in Agenda 21 and other international agreements.
- Support sustainable development in nonindustrialized nations.

CH27-25

❖ Models of Sustainable Development

- The Green Party has been instrumental in encouraging sustainable practices in Germany and other countries.
- These countries can serve as models for other nations.
- Taking the lead in sustainable development positions nations to become leading exporters of knowledge and technologies.

CH27-26

27.5 Global Government: Toward a Sustainable World Community

- Many environmental problems threaten the national security of nations and are global in scale.
- Because of this, they will require international cooperation.

CH27-27

Notes

❖ Regional and Global Alliances

- The task of building a sustainable future is global in nature.
- Regional and global alliances are needed to solve the problems that afflict many nations.

CH27-28

❖ Regional and Global Alliances (cont.)

- The Climate Convention is a first step in reducing global greenhouse gases.
- Success has been limited and more is needed to solve the problem of global warming.
- The Biodiversity Convention calls on nations to take steps to protect species.

CH27-29

❖ Regional and Global Alliances (cont.)

- Agenda 21 is a list, containing an estimated 4500 actions, to promote sustainability.
- Despite its size, the document misses very important elements of sustainability, such as:
  - population stabilization
  - the shift to renewable energy

CH27-30

❖ Regional and Global Alliances (cont.)

- By most accounts, the forest principles articulated at the Earth Summit did little to advance global efforts in sustainable forest harvest and management.
- The Rio Declaration espoused political principles essential to the goal of sustainable global development.

CH27-31

❖ Strengthening International Government

- Two schools of thought exist among proponents of strengthening international environmental protection and sustainable development.
  - One advocates a strengthening of existing structures such as the United Nations.
  - The other promotes the creation of a global government.

CH27-32

❖ Strengthening International Government (cont.)

- Several agencies within the United Nations already work toward the goals of environmental protection and sustainable development.
- Strengthening these organizations could greatly improve these efforts.

CH27-33

Notes

Notes

❖ Strengthening International Government (cont.)

- Some people think we need a world government to deal with the many complex global issues confronting human society.

CH27-34

Photo Credits

Slide #	Credit
CH01-5	Courtesy of Scott Bauer/USDA
CH02-2	© Photodisc/Getty Images
CH02-4	© Photos.com
CH02-14 top	Courtesy of Department of Energy
CH02-14 bottom	© Ali Azhar/ShutterStock, Inc.
CH03-4	© AbleStock
CH03-13	© Kent Knudson/PhotoLink/ Photodisc/Getty Images
CH03-14	©AbleStock
CH04-3	© AbleStock
CH04-4	© Photos.com
CH04-7 top left	© AbleStock
CH04-7 top right	Courtesy of U.S. Fish & Wildlife Services
CH04-7 bottom left	© AbleStock
CH04-7 bottom middle	Courtesy of Tim McCabe/NRCS
CH04-7 bottom right	© Photos.com
CH05-2	© AbleStock
CH005-8 left	Courtesy of U.S. Fish & Wildlife Services
CH005-8 right	© Photodisc
CH05-09	© Photos.com
CH05-10	© Photos.com
CH05-11	Courtesy of Tim McCabe/NRCS
CH05-12 top	© AbleStock
CH05-12 bottom	© Photos.com
CH05-13 top	© Photodisc
CH05-13 bottom	© Photodisc
CH05-14	©Tomas Kopecny/Alamy Images
CH05-17	© AbleStock
CH05-19	© AbleStock
CH05-20	Courtesy of National Estuarine Research Reserve Collection
CH05-22 left	© AbleStock
CH05-22 right	Courtesy of NOAA
CH06-2	© Photos.com
CH06-9	Dan Chiras
CH06-11	Courtesy of U.S. Army Corps of Engineers
CH06-12 top	Courtesy of U.S. Army Corps of Engineers
CH06-12 bottom	Courtesy of Carl Thornber/USGS
CH06-17	© Steffen Foerster/ Shutterstock, Inc.
CH06-19	© AbleStock
CH06-20 left	© Photos.com
CH06-20 right	© AbleStock
CH06-22	© AbleStock
CH06-23	© Photos.com
CH06-25	© Kim Pin Tan/ShutterStock, Inc.
CH06-27	© Photodisc
CH06-28	Courtesy of U.S. Fish & Wildlife Services
CH06-30	Courtesy of Gary Kramer/NRCS
CH07-6	©Jack Dagley/Shutterstock, Inc.
CH07-7	© AbleStock
CH08-4	Courtesy of Christine McKeen
CH08-7	© Corbis
CH08-19	© AbleStock
CH10-3	Courtesy of Dr. Lyle Conrad/CDC
CH10-6	© AbleStock
CH10-9	Courtesy of Lynn Betts/NRCS
CH10-10	© AbleStock
CH10-12	Courtesy of Erwin C. Cole/NRCS
CH10-21	©Ken Hammond/USDA
CH10-22	Courtesy of Tim McCabe/ USDA ARS
CH10-23	©Kim Pan Tin/ShutterStock, Inc.
CH10-24	Courtesy of Erwin Cole/NRCS
CH10-26	Courtesy of Lynn Betts/NRCS
CH10-28	© Adrian T. Jones/ ShutterStock, Inc.
CH10-35	Courtesy of National Cancer Institute
CH10-36	© WizData, Inc./ShutterStock, Inc.
CH11-2	Courtesy of the USFWS
CH11-3	Courtesy of SFWMD
CH11-4	Courtesy of NOAA
CH11-8	© WizData, Inc./ShutterStock, Inc.
CH11-9	© AbleStock
CH11-12	Courtesy of NOAA
CH11-14	Courtesy of the U.S. Fish and Wildlife Services
CH11-15	© AbleStock
CH11-17	Courtesy of the NOAA
CH11-18	Courtesy of the USFWS
CH12-3	Courtesy of Tim McCabe/NRCS
CH12-7	© WizData, Inc./ShutterStock, Inc.
CH12-8	Courtesy of the NRCS
CH12-13	Courtesy Jacques Descloitres, MODIS Rapid Response Team, NASA-Goddard Space Flight Center
CH12-16	©Photodisc/Getty Images
CH12-25	©Photos.com
CH12-27	Courtesy of the National Archives and Records Administration
CH13-9	Courtesy USGS Eros Data Center, based on data provided by the Landsat science team.
CH13-11	Courtesy of the U.S. Army Corps of Engineers
CH13-15	© Theunis Jacobus Botha/Shutter-Stock, Inc.
CH13-18	Courtesy of FEMA
CH13-19	Courtesy of FEMA
CH13-18	Courtesy of USFWS
CH13-22	Courtesy of GSFC/NASA
CH13-23	Courtesy of USFWS
CH13-29	Courtesy of Hope Alexander, EPA Documerica/NOAA
CH14-13	Courtesy of D. Hardesty/USGS
CH14-23	© LiquidLibrary
CH15-11	© Ali Azhar/ShutterStock, Inc.
CH15-22	© Photos.com
CH17-5	© T. O'Keefe/PhotoLink/Photodisc/ Getty Images
CH17-6	Courtesy of the Dept. of Energy

CH18-23 . . . . © AbleStock
CH19-10 . . . . ©Chad Littlejohn/ Shutterstock, Inc.
CH19-11 . . . . ©Dean D. Fetterolf/ Shutterstock, Inc.
CH20-5 . . . . © AbleStock
CH20-9 three right figs. . Courtesy of NASA/JPL/Agency for Aerospace Programs (Netherlands)
CH20-14 . . . . © AbleStock
CH20-17 left. . . . . Courtesy of R. Ross Gilchrist/ Copyright © 1999 Weyerhauser. All Rights Reserved.
CH20-17 right. . . . . © isifa Image Service s.r.o./ Alamy Images
CH20-25 . . . . © Marin Siepmann/imagebroker/ Alamy Images
CH20-36 . . . . © Thierry Dagnelie/ Shutterstock, Inc.
CH21-19 . . . . Courtesy of Exxon Valdez Oil Spill Trustee Council/NOAA
CH21-23 . . . . © T. O'Keefe/Photodisc/ Getty Images
CH21-25 . . . . Image courtesy NASA/GSFC/ LaRC/JPL, MISR Team
CH21-26 . . . . Courtesy of NOAA
CH22-18 . . . . Courtesy of Tim McCabe/NRCS
CH22-20 left. . . . . Courtesy of Alan Fletcher Research Station, Queensland Department of Natural Resources and Mines, Australia
CH22-20 right. . . . . Courtesy of Alan Fletcher Research Station, Queensland Department of Natural Resources and Mines, Australia
CH22-23 . . . . Courtesy of Lynn Betts/NRCS
CH23-2 . . . . Courtesy of Bill Empson/Pictures/ U.S. Army Corps of Engineers
CH23-20 . . . . Courtesy of the U.S. Department of Energy
CH24-2 . . . . © Photodisc/Getty Images
CH24-13 . . . . © Photodisc/Getty Images
CH25-12 . . . . Courtesy of Exxon Valdez Oil Spill Trustee Council/NOAA
CH27-6 . . . . Courtesy of Mark Wolfe/FEMA